XLVI Symposium on Nuclear Physics 2025

EPJ Web of Conferences Volume 333 (2025)

Hacienda Cocoyoc, Morelos, Mexico
6-9 January 2025

Editors:

Roelof Bijker
Araceli García Flores
Óscar Eduardo López López
Daniel José Marín Lámbarri
Javier Mas Ruiz
Tochtli Cuauhtli Yépez Martínez

ISBN: 979-8-3313-2496-4

Printed from e-media with permission by:

Curran Associates, Inc.
57 Morehouse Lane
Red Hook, NY 12571

Some format issues inherent in the e-media version may also appear in this print version.

For additional information, please contact EDP Sciences – Web of Conferences
at the address below.

EDP Sciences – Web of Conferences
17, Avenue du Hoggar
Parc d'Activité de Courtabœuf
BP 112
F-91944 Les Ulis Cedex A
France

Phone: +33 (0) 1 69 18 75 75
Fax: +33 (0) 1 69 28 84 91

contact-edps@webofconferences.org

Additional copies of this publication are available from:

Curran Associates, Inc.
57 Morehouse Lane
Red Hook, NY 12571 USA
Phone: 845-758-0400
Fax: 845-758-2633
Email: curran@proceedings.com
Web: www.proceedings.com

TABLE OF CONTENTS

EPJ Web of Conferences **333**, 00001 (2025) https://doi.org/10.1051/epjconf/202533300001
XLVI Symposium on Nuclear Physics 2025

XLVI Symposium on Nuclear Physics
Cocoyoc, Morelos, México.

XLVI Symposium on Nuclear Physics, January 6-9, 2025

The 46th Symposium on Nuclear Physics was celebrated from January 6-9, 2025 at the Hacienda Cocoyoc, Morelos, Mexico. This symposium has been held every year since its inauguration in 1978 with the exception of the COVID years 2021 and 2022.

There was a tremendous response from the international community of nuclear physics which led to a very successful and lively meeting. The program consisted of 37 plenary talks and 38 posters over a broad range of hot topics, such as nuclear structure and reactions, fundamental symmetries and hadron physics, facilities, instrumentation and applications.

These proceedings contain 22 articles of exciting new results that were presented at the meeting.

Roelof Bijker
Instituto de Ciencias Nucleares, UNAM

Araceli García Flores
Instituto de Física, UNAM

Oscar Eduardo López López
Instituto de Física, UNAM

Daniel José Marín Lámbarri
Instituto de Física, UNAM

Javier Mas Ruiz
Instituto de Ciencias Nucleares, UNAM

Tochtli Cuauhtli Yépez Martínez
Instituto de Educación Media Superior, CDMX

International Advisory Committee

- Ani Aprahamian, Notre Dame University, USA
- Daniel Bardayan, Notre Dame University, USA
- Carlos Bertulani, Texas A&M, USA
- Francesco Cappuzzello, INFN-Laboratori Nazionali del Sud, Italy
- Osvaldo Civitarese, Universidad Nacional de La Plata, Argentina
- Martin Freer, University of Birmingham, UK
- Alfredo Galindo-Uribarri, Oak Ridge National Laboratory, USA
- María José García Borge, CSIC, Spain
- David Jenkins, University of York, UK
- Reiner Krücken, Lawrence Berkeley National Laboratory, USA
- Jorge López, The University of Texas at El Paso, USA
- Nilberto Medina, Universidade de São Paulo, Brazil
- Thomas Papenbrock, University of Tennessee, USA
- Jorge Piekarewicz, Florida State University, USA
- Norbert Pietralla, T.U. Darmstadt, Germany
- Rosario Gianluca Pizzone, INFN-Laboratori Nazionali del Sud, Italy
- W. Michael Snow, Indiana University, USA
- Adam Szczepaniak, Indiana University, USA

Statement of Peer review

In submitting conference proceedings to *Web of Conferences*, the editors of the proceedings certify to the Publisher that

1. They adhere to its **Policy on Publishing Integrity** in order to safeguard good scientific practice in publishing.
2. All articles have been subjected to peer review administered by the proceedings editors.
3. Reviews have been conducted by expert referees, who have been requested to provide unbiased and constructive comments aimed, whenever possible, at improving the work.
4. Proceedings editors have taken all reasonable steps to ensure the quality of the materials they publish and their decision to accept or reject a paper for publication has been based only on the merits of the work and the relevance to the journal.

Title, date and place of the conference

XLVI Symposium on Nuclear Physics
January 6-9, Cocoyoc, Mexico

Proceedings editor(s):

Roelof Bijker
Araceli García Flores
Óscar Eduardo López López
Daniel José Marín Lámbarri
Javier Mas Ruiz
Tochtli Cuauhtli Yépez Martínez

Date and editor's signature

July 3, 2025

17, avenue du Hoggar - PA de Courtabœuf – BP 112 - 91944 Les Ulis Cedex A (France)
Tél. : 33 (0)1 69 18 75 75 - Fax : 33(0)1 69 07 45 17 – www.edpsciences.org

https://doi.org/10.1051/epjconf/202533301001

Strong Electromagnetic Fields in Heavy Ion Collisions

C.A. Bertulani[1,2,3,*]

[1]Department of Physics and Astronomy, East Texas A&M University, Commerce, Texas, 75025, USA
[2]ExtreMe Matter Institute EMMI, GSI, Darmstadt, Germany
[3]Institut für Kernphysik, Technische Universität Darmstadt, Germany

Abstract. Photons play a pivotal role in both theoretical and applied physics, ranging from elementary particle studies to practical technologies. This article explores how ultra-peripheral collisions of heavy ions act as sources of nearly real photons, enabling the investigation of a wide range of photon-induced processes. By applying the equivalent photon approximation, one can study nuclear structure, QED phenomena, and potential new physics, all in a background-suppressed environment.

1 Virtual and real photons

The study of electromagnetic processes in high-energy nuclear physics has benefited significantly from the development of the Equivalent Photon Method (EPM), which originated from an idea proposed by Enrico Fermi in 1924 [1]. His pioneering approach suggested that the time-dependent electromagnetic field of a fast-moving charged particle could be represented as a flux of quasi-real photons. This concept, though initially classical, has been refined using the language of quantum electrodynamics (QED) and perturbation theory, making it applicable to a wide range of processes involving nuclei and particles.

In the context of ultra-relativistic heavy ion collisions, such as those occurring at the Relativistic Heavy Ion Collider (RHIC) or the Large Hadron Collider (LHC), the Lorentz-contracted electromagnetic fields of the nuclei become extremely intense. These fields interact at large impact parameters, allowing the nuclei to remain intact while exchanging energy and momentum via virtual photon interactions. This regime, known as ultra-peripheral collisions (UPCs), is ideal for applying the EPM because the electromagnetic interaction dominates over hadronic contributions, which are suppressed by the large distance between the colliding nuclei.

The central premise of the EPM is to replace the time-dependent electromagnetic field with a distribution of equivalent photons. The interaction of these photons with a target can then be treated as if the photons were real. For each electric or magnetic multipolarity of the interaction, denoted by E or M, and characterized by angular momentum L, one defines an equivalent photon number density $n_{E/M,L}(\omega)$ as a function of the photon energy ω. The total cross section for a process can then be expressed as [2, 3]:

$$\sigma = \sum_{E/M,L} \int \frac{d\omega}{\omega} n_{E/M,L}(\omega) \sigma_\gamma^{(E/M,L)}(\omega), \tag{1}$$

*e-mail: carlos.bertulani@etamu.edu

where $\sigma_\gamma^{(E/M,L)}(\omega)$ is the cross section for the real photon process at energy ω.

The photon number $n_{E/M,L}(\omega)$ encapsulates the kinematic and dynamic aspects of the colliding system. It depends on the Lorentz factor γ of the projectile, the impact parameter b, the nuclear charge Z, and the transition multipolarity. For practical calculations, one integrates over all possible impact parameters larger than a minimum value $b_{\min}$ to exclude overlapping configurations where hadronic interactions may occur. The result is a spectrum of equivalent photons that can be used to compute cross sections for a wide variety of processes.

A key advantage of this approach is that it allows one to leverage tabulated or experimentally measured photoabsorption cross sections for processes such as nuclear excitation, particle production, and meson photoproduction. It also enables theoretical predictions based on model-dependent photonuclear reaction mechanisms. Examples include dipole excitations in stable and exotic nuclei, Coulomb excitation of low-lying collective states, and the production of vector mesons in coherent photon-nucleus interactions. It plays a critical role in understanding photon-photon and photon-gluon fusion mechanisms, contributing to our knowledge of QCD dynamics.

Attention is often directed toward the regime of high energies, where the interaction cross section can be approximated by

$$\sigma \simeq \sum_{E/M,L} \int \frac{d\omega}{\omega} n(\omega) \sigma_\gamma^{(E/M,L)}(\omega) = \int \frac{d\omega}{\omega} n(\omega) \sum_{E/M,L} \sigma_\gamma^{(E/M,L)}(\omega) = \int \frac{d\omega}{\omega} n(\omega) \sigma_\gamma(\omega), \quad (2)$$

with the total real-photon cross section defined by $\sigma_\gamma = \sum_{E/M,L} \sigma_\gamma^{(E/M,L)}$.

Equations 1 and 2 are derived under the assumption that only first-order processes contribute [3]. However, the photon is not a simple point-like particle; it can fluctuate into various intermediate states via quantum loop effects. These fluctuations modify its wavefunction, leading to a superposition of bare and virtual components:

$$|\gamma\rangle \;=\; C_{bare} |\gamma_{bare}\rangle + C_{ee} \left|e^- e^+\right\rangle + \cdots + C_{qq} |q\bar{q}\rangle + C_\omega |\omega\rangle + C_\phi |\phi\rangle + C_\rho |\rho\rangle + \cdots \quad (3)$$

Due to its quantum numbers $J^P = 1^-$, the photon readily converts into neutral vector mesons such as ρ, ω, ϕ, or J/ψ through quark-antiquark fluctuations, consistent with the principles of the vector dominance model. Additionally, the photon is inherently multicomponent in energy space, some components are off-shell or virtual. It may undergo processes involving internal splitting, emission of constituent parts, transitions through intermediate hadronic states, or reabsorption cycles [4].

2 Coulomb dissociation and nuclear astrophysics

Coulomb dissociation is an indirect approach to study radiative capture reactions, particularly those of relevance to nuclear astrophysics [5]. In this method, a projectile nucleus with a cluster structure, or a bound nucleon plus a core, is exposed to the intense electromagnetic field of a high-Z target nucleus. The interaction, occurring at large impact parameters to suppress nuclear interactions, mimics photo-disintegration by causing the projectile to break apart. By analyzing the resulting fragments and their energy spectra, one can infer the inverse process cross section using detailed balance principles, i.e., in the reaction $a+A \longrightarrow b+c+A$:

$$\frac{d\sigma_C^{E/M,L}(\omega)}{d\omega d\Omega} = \frac{dn^{E/M,L}(\omega;\theta;\phi)}{d\omega d\Omega} \cdot \sigma_{\gamma+a\to b+c}^{E/M,L}(\omega). \quad (4)$$

Here, $dn^{E/M,L}/d\omega d\Omega$ dependens on the scattering angle $\Omega = (\theta, \phi)$ [5]. Time reversal relates the reaction cross section for $b + c \longrightarrow a + \gamma$ to $\sigma_{\gamma+a\to b+c}^{E/M,L}(\omega)$. This method provides a clean

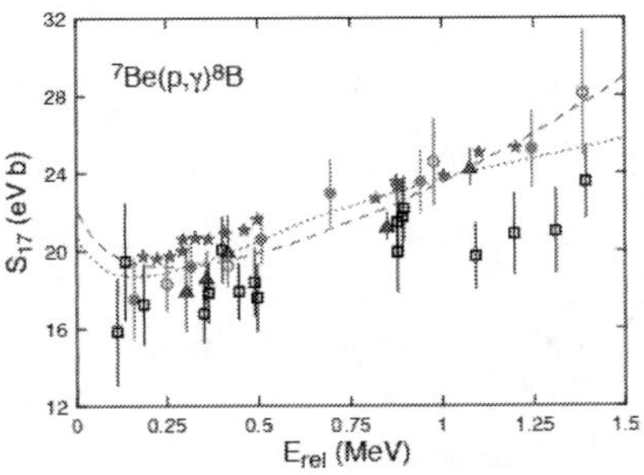

Figure 1. ^{7}Be(p,γ)^{8}B reaction at low energies. Calculations using the no-core shell-model [7] are shown by a dashed-line. The dotted line is a calculation using a resonant group method [8]. Experimental data are also shown [12–19].

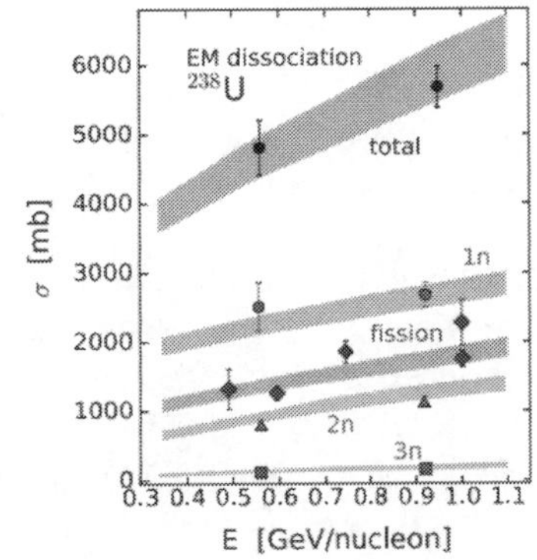

Figure 2. EM dissociation cross sections of ^{238}U projectiles. The projectiles fission (diamonds) and/or emit multiple neutrons (xn) [20–23]. Uncertainty bands for the numerical calculations are shown [24].

probe of neutron or proton capture rates, which are essential in nucleosynthesis pathways like the r-process or rp-process.

Experiments using this method have been successfully carried out at facilities such as RIKEN, GSI, and NSCL, targeting neutron-rich and proton-rich nuclei. Examples include the dissociation of ^{8}B, ^{11}Be, ^{15}C, and ^{19}C, providing critical information on astrophysical S-factors and nuclear halo structures. Due to the purely electromagnetic nature of the interaction, theoretical modeling is more reliable, with fewer uncertainties than hadronic reaction mechanisms. An example of application of this method is the ^{7}Be$(p,\gamma)^8$B reaction [6] as shown in Figure 1. The experimental data displayed as red dots were used the Coulomb dissociation method. See also Refs.[9–11].

The Coulomb dissociation method also serves to validate nuclear structure models, particularly those describing the continuum and resonant states. Its predictive power and ability to provide spectroscopic data on unbound states make it an indispensable tool in both fundamental nuclear physics and applied astrophysics. The contribution of nuclear breakup to the dissociation process has been studied and quantified in many works (see, e.g., Ref. [25]). Loosely-bound systems, or "halo nuclei" with small nucleon separation energies, are subjected to multiple-step higher-order effects including continuum-continuum couplings Ref. [25].

3 Excitation and decay of giant resonances

Giant resonances are collective excitations of the entire nucleus, where nucleons oscillate in a coherent manner under the influence of an external field. These modes are essential to understanding bulk nuclear properties, such as incompressibility, symmetry energy, and the effective interactions among nucleons. Among the most studied are the isovector giant dipole resonance (IVGDR), isoscalar monopole resonance (GMR), and quadrupole resonances.

Ultra-peripheral collisions offer a unique environment to excite giant resonances through the absorption of virtual photons [27, 33, 34]. As the electromagnetic field of a passing ion interacts with the nucleus, it can transfer energy sufficient to trigger these coherent modes. The excitation cross section depends on the strength function of the nucleus and the equivalent photon spectrum. For targets with large charge Z, the nuclear dissociation cross sections are much smaller than electromagnetic [3, 20–23]. In UPCs, their study has provided new insights in studies of nuclear fission [28–30] and as a luminosity monitor in relativistic heavy-ion colliders [31, 32, 39] (see Figure 2).

The IVGDR is particularly sensitive to the neutron-to-proton ratio and is a key observable in determining the symmetry energy of nuclear matter. On the other hand, the GMR provides direct insight into nuclear compressibility, a parameter relevant to both finite nuclei and neutron star matter. Measurement of these modes requires detection of emitted gamma rays or decay particles following resonance excitation. The excitation of the double IVGDR in UPCs was proposed in Refs. [33, 34]. They were identified in experiments carried out at the GSI/Darmstadt laboratory [35, 36] using gamma-gamma coincidences from the decay of the decay of the double GDR (DGDR) [36], as well as by identifying neutron emission from its decay [35] (see Figure 3). The DGDR strength and width provide additional constraints of nuclear models for research on giant resonance [37, 38, 40]. This contributes significantly to the ongoing effort to constrain the nuclear equation of state.

4 Lepton pair production

One of the fundamental processes accessible in ultra-peripheral collisions is the creation of electron-positron pairs from the vacuum. This quantum electrodynamical (QED) phenomenon occurs when two colliding nuclei generate sufficiently strong electromagnetic fields to facilitate two-photon interactions. The process is described by lowest-order Feynman diagrams in QED and becomes increasingly probable with rising nuclear charge and beam energy. In its simplest form, pair production involves two virtual photons fusing to create a e^+e^- pair. The cross section scales approximately with the square of the nuclear charge and logarithmically with the Lorentz factor of the colliding ions, in the form [41–46]

$$\sigma_{e^+e^-} = \frac{28}{27\pi}(Z_1Z_2\alpha r_e)^2 \ln^3\left(\frac{\gamma}{2}\right). \tag{5}$$

The appearance of a triple logarithmic dependence originates from the interplay of several quantum electrodynamical mechanisms: (1) The momentum integration over the exchanged virtual photon between one nucleus and the e^+e^- pair yields a logarithmic term, as $\int dq^2/q^2 \sim \ln(q_{\max}/q_{\min}) \sim \ln\gamma$. (2) Since the produced pair tends to emerge at small angles relative to the beam axis, the angular integration gives another factor, $\int d\theta/\theta \sim \ln\gamma$. (3) The equivalent photon distribution from the opposing nucleus scales as $dN/d\omega \sim 1/\omega$, and the corresponding energy integration $\int_{\omega_{\min}}^{\omega_{\max}} d\omega/\omega \sim \ln\gamma$ introduces the third logarithm. These logarithmic contributions are characteristic features of QED in the ultra-relativistic domain, where boosted fields and collinear emission dominate the dynamics. Higher-order corrections include multiple pair production, Coulomb distortion effects, and screening phenomena are also important [3, 47].

A straightforward model for evaluating multiple pair production was based on the assumption that the number of produced pairs follows a Poisson distribution as a function of impact parameter. In this approach, the probability of generating n e^+e^- pairs at a given b is expressed as [3]

$$P_{e^+e^-}(b) = \frac{[P_0(b)]^n}{n!}\exp\left[-P_0(b)\right], \tag{6}$$

where $P_0(b)$ denotes the single-pair production probability [3], calculated by

$$P_0(b) = \frac{14}{9\pi^2 b^2}(Z_1Z_2\alpha r_e)^2 \ln^2\left(\frac{\gamma}{2m_e b}\right), \tag{7}$$

which applies within the interval $1/m_e < b < \gamma/m_e$ [3]. Several studies have extended this analysis to account for final-state effects in processes involving multiple pairs [48–51].

Measurements performed with the STAR detector at RHIC [52], and later by CMS at CERN [53], provided empirical support for these quantum electrodynamics-based predictions. The formation of other lepton or meson pairs via photon fusion, such as $\gamma\gamma \rightarrow \mu^+\mu^-$, $\gamma\gamma \rightarrow \pi^+\pi^-$, and $\gamma\gamma \rightarrow W^+W^-$, is also relevant, as shown in theoretical results from Refs. [3, 47, 54], and later corroborated [3, 3, 39, 54]. The formulas presented above are applicable to $\mu^+\mu^-$, $\pi^+\pi^-$, and $\tau^+\tau^-$ production under the condition that $\gamma \gg 16$ for muons and $\gamma \gg 200$ for taus. For lower values of γ, modifications to the expressions become necessary [3].

Beyond electrons, heavy lepton pair production and even bound-free pair creation, where the electron is captured by one of the nuclei, have also been observed. These processes contribute to beam losses in accelerators and are of practical relevance in beam lifetime studies.

5 Anti-hydrogen and other exotic atoms

A notable application of ultra-peripheral collisions (UPCs) involves lepton pair production where the negatively charged lepton is subsequently captured by one of the colliding nuclei [3, 56, 57]. The cross section for such a process, specifically, when the electron is trapped in a K-shell orbital, can be approximated by [3]

$$\sigma_K = \frac{33}{20} Z_1^5 Z_2^5 \alpha^5 r_e^2 \ln\left(\frac{\gamma}{2}\right). \quad (8)$$

To account for electron capture into higher atomic shells, one must multiply this result by the factor $\sum_n 1/n^3 \approx 1.202$. For high-$Z$ systems, electron screening and wavefunction distortions alter these values significantly [3].

This mechanism was proposed as a pathway for producing antihydrogen at the LHC [58], and its experimental verification came in a landmark CERN experiment in 1996 [59]. Conducted at the Low Energy Antiproton Ring (LEAR), this effort involved collisions between protons and antiprotons, with the latter capturing positrons to form neutral anti-atoms. This represented the first successful production of antihydrogen in a laboratory setting, with 11 atoms recorded, an achievement widely covered in major news outlets including the New York Times [60].

A similar experiment performed at Fermilab [61] identified 57 antihydrogen events, which were consistent with earlier theoretical forecasts [56, 62]. Since then, anti-atom investigations have continued using ion trapping technologies to probe fundamental symmetries [63, 64]. The utility of UPCs extends beyond antihydrogen: processes leading to anti-deuterium, anti-tritium, and even anti-helium nuclei have been proposed [65]. UPCs also provide a potential source of exotic atoms, such as muonic or pionic atoms, formed via coherent photon exchange, as outlined in [57].

The concept of e^+e^- pair production accompanied by electron capture into atomic orbitals has also been suggested as a source of beam loss in high-energy colliders [3]. Early estimates predicted that beams of heavy ions could degrade within hours due to this effect at the LHC. These predictions have been substantiated by further theoretical analyses [66] and recent experimental confirmations [67].

Another compelling aspect of UPCs is the creation of bound e^+e^- states, positronium, in either ortho or para configurations. Theoretical descriptions for positronium and related bound states, including mesons viewed as $q\bar{q}$ systems, have been developed within the framework of quantum field theory [68]. Fusion of two or three photons in UPCs can give rise to such bound states, as explored in Refs. [69–71]. In total, six leptonic atoms are theoretically possible: (a) positronium (e^+e^-), (b) muonium (μ^+e^-), (c) dimuonium ($\mu^+\mu^-$), (d) tauonium

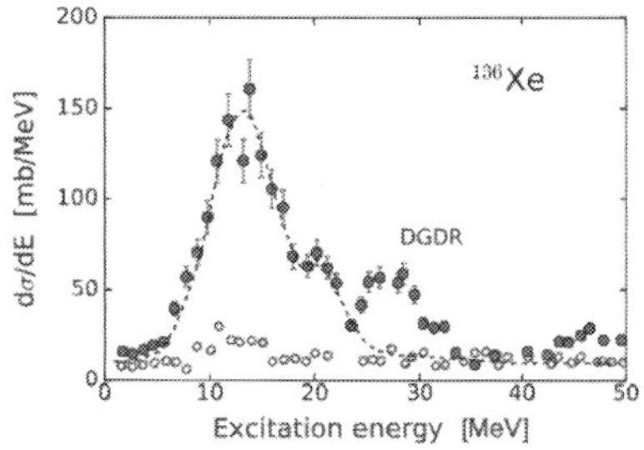

Figure 3. Giant resonances in ^{136}Xe observed in UPCs as a function of the excitation energy. A distinct peak at higher energy displays the Giant Dipole Resonance (DGDR) [35].

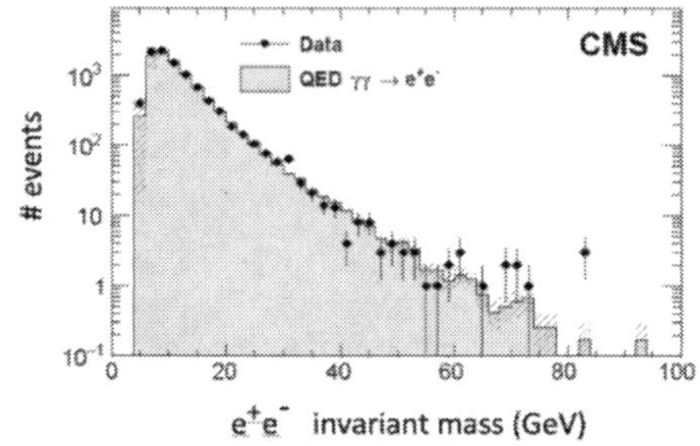

Figure 4. Number of events of e^+e^- pairs created in UPCs obtained with the CMS detector at the LHC (from Ref. [53]).

(τ^+e^-), (e) tau-muonium ($\tau^+\mu^-$), and (f) ditauonium ($\tau^+\tau^-$). Of these, only positronium, muonium, and the positronium molecule (dipositronium) have been observed so far [72–74].

Dimuonium is expected to be significantly more compact than either positronium or muonium, due to the heavier mass of the muons. Its unique mass scale renders it sensitive to potential phenomena beyond the Standard Model [75], particularly in time-like domains of QED. Detecting dimuonium would represent a major milestone in fundamental physics [76]. Production rates for dimuonium in UPC environments at the LHC have been calculated and found to be within measurable reach [77–80]. Moreover, recent work suggests that ditauonium might also be observed under similar conditions [82], with detection strategies relying on its displaced decay vertex and manageable background from dimuon events.

6 Photon-photon fusion

The process of elastic light-by-light scattering, $\gamma + \gamma \longrightarrow \gamma + \gamma$, is mediated through rare quantum fluctuations, where a photon temporarily transforms into a particle-antiparticle pair. Due to the extremely small likelihood of such fluctuations, direct measurements using real photons remain impractical. In Ref. [3], we suggested that ultra-peripheral collisions (UPCs) could serve as an effective method for probing this phenomenon via the reaction $Z_1 + Z_2 \longrightarrow Z_1 + Z_2 + \gamma + \gamma$. In this scenario, two virtual photons interact through a loop (box) diagram, resulting in the emission of two real photons. These calculations, based on Delbrück scattering formalism, involve considerable theoretical uncertainties [3]. The leading-order cross section for $\gamma^*\gamma^* \longrightarrow \gamma\gamma$ in UPCs, particularly for high-energy photon final states, is given by [3]

$$\sigma_D \sim 2.54 \times 10^{-2} Z^4 \alpha^4 r_e^2 \ln^3\left(\frac{\gamma}{m_e R}\right), \tag{9}$$

where R denotes the radius of the heavy-ion nucleus. For Pb+Pb collisions at the LHC, the resulting cross section can be quite large, on the order of $\sigma_D \sim 50$ b. However, not all outgoing photons can be attributed unambiguously to this channel, as backgrounds, such as those from Bremsstrahlung, complicate the signal extraction. For photon energies $E_\gamma > m_\mu$, the cross section becomes significantly smaller, approximately $\sigma_D \sim 30$ nb [81].

Further theoretical studies expanded on these ideas, introducing alternative photon-photon scattering mechanisms including the VDM-Regge framework, two-gluon exchange models, and meson resonance contributions (see, for example, Refs. [83–87]). A major experimental milestone came with the ATLAS detector's observation of this elusive process at the LHC (Figure 5) [88], lending credibility to prior theoretical models and opening the door to probing scenarios beyond the Standard Model (SM). Any measured cross section that exceeds the SM expectation could hint at the presence of new fundamental particles, such as axions [89–91].

7 Production of Mesons

To extend the Equivalent Photon Number (EPN) framework to include the formation of bound states in ultra-peripheral collisions, we proposed a method for calculating the production cross section of a generic particle X [3]. The photon-photon fusion cross section can be written as

$$\sigma_X = \int \frac{d\omega_1}{\omega_1}\frac{d\omega_2}{\omega_2} n_\gamma(\omega_1)\, n_\gamma(\omega_2)\, \sigma^X_{\gamma\gamma}(\omega_1\omega_2), \tag{10}$$

where $n_\gamma(\omega)$ represents the equivalent photon flux, and $\sigma^X_{\gamma\gamma}(\omega_1\omega_2)$ is the cross section for producing particle X via two-photon interactions. The latter can be determined through a result due to Low [92], which connects the formation cross section with the particle's two-photon decay width:

$$\sigma^X_{\gamma\gamma}(\omega_1\omega_2) = 8\pi^2(2J+1)\frac{\Gamma_{m_X\to\gamma\gamma}}{m_X}\,\delta\left(\omega_1\omega_2 - m_X^2\right). \tag{11}$$

Here, m_X and J represent the mass and spin of the final state, while $\Gamma_{m_X\to\gamma\gamma}$ denotes its two-photon decay width. A more comprehensive analysis of mesonic and exotic candidates for such processes, including states often omitted from standard treatments, was presented in Ref. [93].

A worthy application of this formalism was proposed in 1989, aimed at identifying the Higgs boson in ultra-peripheral LHC collisions [94]. Using Equations (10, 11) and assuming plausible Higgs parameters, we estimated a production cross section of approximately 1 nb [3]. This value is on par with Higgs production in hadronic processes but with significantly reduced background particle production. Later studies pointed out that direct $\gamma\gamma \to b\bar{b}$ processes exhibit substantially larger cross sections [95], presenting a serious challenge for isolating the Higgs signal. A more favorable scenarios for Higgs detection in UPCs have been discussed in subsequent works [96, 97]. Since the Higgs predominantly decays into $b\bar{b}$ pairs, distinguishing it from this background remains a major challenge. Ultimately, the Higgs boson was confirmed via hadronic channels at the LHC by the ATLAS and CMS collaborations [98, 99].

Exotic hadronic configurations such as multiquark states, including molecular structures like $(q\bar{q})(q\bar{q})$, glueballs composed of gluons (gg), and hybrid mesons involving a gluon $(q\bar{q}g)$, play a pivotal role in understanding meson spectroscopy [100]. Ultra-peripheral collisions (UPCs) provide a promising experimental setting to explore these unconventional resonances via their unusual $\gamma\gamma$ couplings and characteristic energy levels. Such investigations can test theoretical expectations regarding non-standard states [68–70].

The decay width for $\gamma\gamma$ interactions is particularly informative, as it reflects the net electric charge of the quark constituents. This property can help distinguish between quark-based resonances and glueball candidates, which lack valence quarks. Therefore, the absence of certain meson states in $\gamma\gamma$ fusion experiments may serve as strong evidence for their gluonic origin [68–70]. In UPC scenarios, glueballs are typically produced through quark-antiquark annihilation into gluon pairs, whereas conventional mesons with $q\bar{q}$ structure are generated through direct photon-photon interactions.

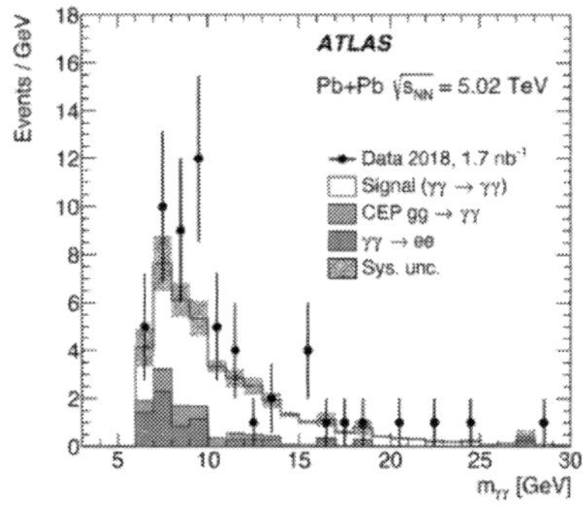

Figure 5. Light-by-light events obtained at the ATLAS detector [88].

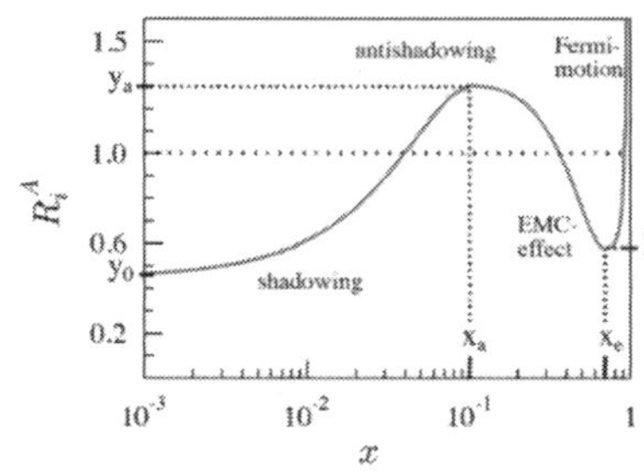

Figure 6. Medium effect dependence with the momentum fraction x.

8 Parton distribution functions

The study of vector meson production, such as for J/ψ and $\Upsilon(1s)$, can be addressed using Eq. (1), and early theoretical and experimental investigations were reported in Refs. [101–106]. In 2001, it was proposed [107] that coherent photoproduction of vector mesons could be utilized to probe the generalized parton distributions (GPDs) in nuclei, denoted $F_A(x, Q^2)$. This approach builds upon the formalism developed in Ref. [108], with the production cross section near zero momentum transfer given by

$$\left.\frac{d\sigma^{\gamma A\to VA}}{dt}\right|_{t=0} = \frac{16\pi\alpha_s^2(Q^2)\Gamma_{ee}}{3\alpha M_V^5}\left[xF_A(x,Q^2)\right]^2, \tag{12}$$

where $\alpha_s(Q^2)$ is the strong coupling constant evaluated at a pQCD scale $Q^2 = W_{\gamma g}^2$. Here, M_V is the mass of the produced vector meson, Γ_{ee} is its partial width for decay into e^+e^-, and $x = M_V^2/W_{\gamma p}^2$ corresponds to the gluon momentum fraction.

The nuclear parton distribution function (PDF), $F_a^A(\mathbf{r}, x, Q^2)$, describes the spatial and momentum distribution of a parton species a within a nucleus A, and can be represented as a convolution:

$$F_a^A(\mathbf{r}, x, Q^2) = R_a^A(\mathbf{r}, x, Q^2)\cdot f_a(x, Q^2),$$

where R_a^A encodes the nuclear modification, and $f_a(x, Q^2)$ denotes the parton distribution in a free nucleon [109]. The spatial coordinate $\mathbf{r}$ identifies the nucleon's position inside the nucleus. Different regions of Bjorken-x exhibit distinct nuclear effects. At small values ($x < 0.04$), parton densities are reduced relative to the nucleon baseline, a phenomenon known as shadowing, where $R_a^A < 1$ (see Figure 6). In the intermediate range $0.04 < x < 0.3$, the opposite behavior—antishadowing—occurs with $R_a^A > 1$. Between $0.3 < x < 0.8$, the EMC effect is dominant, and at large $x > 0.8$, enhancement arises due to nucleon Fermi motion. Each of these modifications reflects different physical mechanisms affecting parton distributions in the nuclear environment. In Ref. [109], we analyzed how different gluon parton distribution functions (PDFs) affect the production of vector mesons such as J/ψ and $\Upsilon(1s)$ in ultra-peripheral collisions. In particular, we distinguished between two dominant mechanisms in UPCs involving pPb and PbPb systems: the "direct" and "resolved" channels. The direct mechanism involves a photon interacting coherently with the nucleus, while the resolved process involves the photon fluctuating into a partonic configuration, typically a $q\bar{q}$ pair, which subsequently scatters off the nucleus.

At leading order, the direct channel is controlled by the gluon content of the nucleus, which is poorly constrained at small momentum fractions x (see Figure 8). In contrast, the

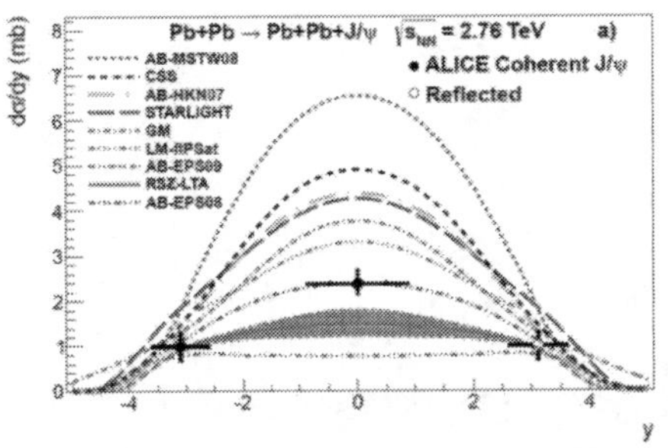

Figure 7. Rapidity y distribution of J/Ψ production in UPCs at the LHC. Different Parton Distribution Functions (PDFs) are tested [110].

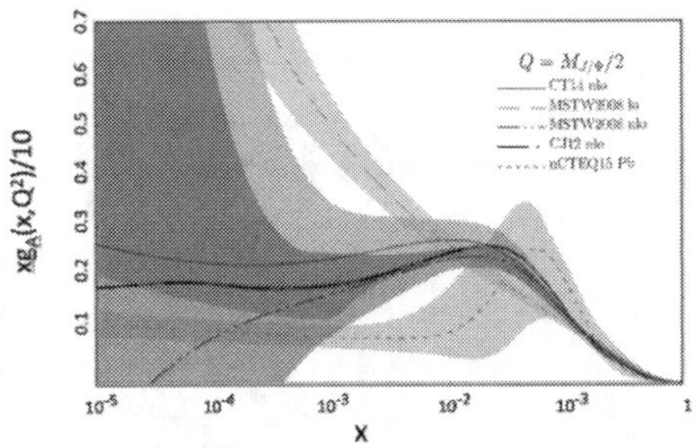

Figure 8. Compilation of gluon distribution functions depending on the momentum fraction x. Large error bands are evident.

resolved channel is sensitive to parton distributions in both the nucleus and the photon, including light quarks and gluons. Since vector meson production in UPCs depends quadratically on the gluon density, these processes provide enhanced sensitivity to the underlying nuclear gluon distributions, impacting both total and differential observables such as rapidity spectra [107, 109]. Our earliest predictions for J/ψ production, using nuclear gluon PDFs that incorporate shadowing effects [109], were found to be in excellent agreement with measurements performed by the ALICE collaboration [110–113], as shown in Figure 7. These findings support the use of J/ψ and Υ photoproduction in UPCs as a sensitive tool for probing gluon shadowing at very small x values, specifically in the regime $x < 10^{-3}$.

Currently, a widely supported view is that the resolved photon components play a crucial role in vector meson photoproduction within UPCs. Multiple studies [114–118] have emphasized the significant impact of hadronic fluctuations of the photon, especially in light meson production, and the relevance of leading-twist nuclear shadowing in $q\bar{q}$ photoproduction off heavy nuclei.

9 Perspectives

Ideas that first emerged during the 1980s and 1990s have since matured into a rich area of exploration at modern relativistic colliders. Ultra-peripheral collisions (UPCs) have revealed an array of surprising phenomena, such as the excitation of double giant resonances, production and investigation of anti-atoms, beam losses resulting from electron capture, light-by-light scattering, constraints on parton distribution functions (PDFs), and opportunities to probe physics beyond the Standard Model. These collisions could, in principle, enable the detection of processes like $\gamma\gamma \to$ graviton [119] or the production of axion-like particles [89, 90]. Whether such discoveries are feasible remains an open question, dependent on advancements in beam dynamics, accelerator performance, and innovative detector technologies.

Should searches at CERN continue to yield no evidence for supersymmetric particles, the focus may shift toward alternative strategies. These include refining precision calculations and experimental measurements, probing unconventional mesonic states, or further investigating how the nuclear medium influences parton distributions within hadrons.

Acknowledgments. This work was partially supported by the U.S. DOE grant DE-FG02-08ER41533, and the ExtreMe Matter Institute EMMI at the GSI Helmholtzzentrum für Schwerionenforschung, Darmstadt, Germany.

References

[1] E. Fermi, Zeit. Phys. **29**, 315 (1924); E. Fermi, Nuovo Cimento **2**, 143 (1925).
[2] C.A. Bertulani and G. Baur, Nucl. Phys. A **442**, 739 (1985).
[3] G. Baur and C.A. Bertulani, Physics Reports **163**, 299 (1988).
[4] C.A. Bertulani, S. Klein and J. Nystrand, Annu. Rev. Nucl. Part. Sci. **55**, 271 (2005).
[5] G. Baur, C. Bertulani and H. Rebel, Nucl. Phys. A **459**, 188 (1986).
[6] T. Motobayashi, et al., Phys. Rev. Lett. **73**, 2680 (1994).
[7] P. Navrátil, C.A. Bertulani, E. Caurier, Phys. Lett. B **634**, 191 (2006); PRC **73**, 065801 (2006).
[8] P. Descouvemont, D. Baye, Nuclear Phys. A **567**, 341 (1994).
[9] H. Esbensen, G.F. Bertsch, and K. Snover, Phys. Rev. Lett. **94**, 042502 (2005).
[10] M. Gai, Phys. Rev. Lett. **96**, 159201 (2006).
[11] E.G. Adelberger et al., Rev. Mod. Phys. **83**, 195 (2011).
[12] F.J. Vaughn, R.A. Chalmers, D. Kohler, L.F. Chase, Phys. Rev. C **2**, 1657 (1970).
[13] B.W. Filippone, A.J. Elwyn, C.N. Davids, D.D. Koetke, Phys. Rev. C **28**, 2222 (1983).
[14] L.T. Baby, et al., Phys. Rev. Lett. **90**, 022501 (2003).
[15] A.R. Junghans, et al., Phys. Rev. C **68**, 065803 (2003).
[16] N. Iwasa, et al., Phys. Rev. Lett. **83**, 2910 (1999).
[17] B.Davids, et al., Phys. Rev. Lett. **86**, 2750 (2001).
[18] F. Schümann, et al, Phys. Rev. Lett. **90**, 232501 (2003).
[19] R.W. Kavanagh, T.A. Tombrello, et al., Bull. Am. Phys. Soc. **14**, 1209 (169).
[20] S. Polikanov, et al., Z. Phys. A **350**, 221 (1994).
[21] P. Armbruster, et al., Z. Phys. A **355**, 191 (1996).
[22] T. Rubehn, et al., Z. Phys. A **353**, 197 (1995).
[23] T. Aumann, et al., Nuclear Phys. A **599**, 321 (1996).
[24] T. Aumann, C.A. Bertulani and K. Suemmerer, Phys. Rev. C **51**, 416 (1995).
[25] C.A. Bertulani, Phys. Rev. C **108**, 054602 (2023).
[26] T. Aumann, K. Suemmerer, C.A. Bertulani and J.V. Kratz, Nucl. Phys. A **599**, 321 (1996).
[27] G. Baur and C.A. Bertulani, Nucl. Phys. A **458**, 725 (1986).
[28] T. Aumann, K. Boretzky, J. Stroth, et al., Acta Physica Polonica. B **1**, 375 (1997).
[29] C.A. Bertulani, Y. Kucuk and R. Lozeva, Phys. Rev. Lett. **124**, 132301 (2020).
[30] I. Stetcu, et al., Phys. Rev. Lett. **114**, 012701 (2015).
[31] A.J. Baltz, C. Chasman and S.N. White Nucl. Inst. Meth. Phys. Res. A **41**, 1 (1998).
[32] Spencer R. Klein, Nucl. Inst. meth. Phys. **459**, 1 (2001).
[33] G. Baur and C.A. Bertulani, Phys. Lett. B **174**, 23 (1986).
[34] G. Baur and C.A. Bertulani, Phys. Rev. C **34**, 1654 (1986).
[35] R. Schmidt, et al., Phys. Rev. Lett. **70**, 1767 (1993).
[36] J. Ritman, et al Phys. Rev. Lett. **70**, 533 (1993).
[37] H. Emling, Prog. Part. Nucl. Phys. **33**, 729 (1994).
[38] T. Aumann, P.F. Bortignon and H. Emling, Ann. Rev. Nucl. Part. Sci. **48**, 351 (1998).
[39] C.A. Bertulani and G. Baur, Physics Today **47**, 22 (1994).
[40] C.A. Bertulani and V. Ponomarev, Phys. Reports **321**, 139 (1999).
[41] W.H. Furry and J.F. Carlson, Phys. Rev. **44**, 238 (1933).
[42] H.A. Bethe, Proc. Cambridge Philos. Soc. **30**, 524 (1934).
[43] H.J. Bhabha, Proc. R. Soc. London Ser. A **152**, 559 (1935).
[44] Y. Nishina, S. Tomonaga and H. Tamaki, Sci. Pap. Phys. Math. Res. Jpn **24**, 137 (1934).
[45] Y. Nishina, S. Tomonaga and M. Kobayashi, Sci. Pap. Inst. Phys. Chem. Res. **27**, 137 (1935).
[46] G. Racah, Nuovo Cimento **14**, 93 (1937).
[47] G. Baur and C.A. Bertulani, Nucl. Phys. A **505**, 835 (1989).
[48] G. Baur, K. Hencken, D. Trautmann, S. Sadovsky, Y. Kharlov, Phys. Reports **364**, 359 (2002).

[49] K. Hencken, D. Trautmann, G. Baur, Phys. Rev. A **51**, 998 (1995).
[50] G. Baur, K. Hencken and D. Trautmann Phys. Reports **453**, 1 (2007).
[51] A.J. Baltz et al. Phys. Reports **458**, 1 (2008).
[52] J. Adams et al. (STAR Collaboration), Phys. Rev. C **70**, 031902 (2004).
[53] W. Adam, et al., Phys. Lett. B **797**, 303 (2008).
[54] G. Baur and C.A. Bertulani, Phys. Rev. C **35**, 836 (1987).
[55] C.A. Bertulani and G. Baur, Z. Phys. A **330**, 77 (1988).
[56] C.A. Bertulani C A and G. Baur, Phys. Rev. D **58** 034005 (1998).
[57] C.A. Bertulani and M. Ellermann, Phys. Rev. C **81** 044910 (2010).
[58] C. Munger and S. Brodsky, Phys. Rev. D **49**, 3228 (1994).
[59] G. Baur, et al, Phys. Lett. B **368**, 3 (1996).
[60] Malcolm W. Browne, The New York Times, January 5 (1996).
[61] G. Blanford et al., Phys. Rev. Lett. **80**, 3037 (1998).
[62] C.A. Bertulani and D. Dolci, Nucl. Phys. A **683**, 635 (2001).
[63] E.S. Reich, Nature News **11**, 17 (2010).
[64] L. Grossman Phys. Rev. Focus **26** (2010).
[65] H. Agakishiev, et al, Nature, **473**, 353 (2011).
[66] S.R. Klein, Phys. Rev.: Accelerators and Beams **17**, 121003 (2014).
[67] J.M. Schaumann, et al. , Phys. Rev.: Accelerators and Beams **23**, 121003 (2020).
[68] C.A. Bertulani and F.S. Navarra Nucl. Phys. A **703**, 861 (2002).
[69] B.D. Moreira, C.A. Bertulani, V.P. Goncalves and F. S. Navarra Phys. Rev. D **94**, 094024 (2016).
[70] C.A. Bertulani, V.P. Goncalves, B.D. Moreira and F.S. Navarra Eur. Phys. J. **137**, 06019 (2017).
[71] R. Fariello, D. Bhandari, C. A. Bertulani and F. S. Navarra, Phys. Rev. C **108**, 044901 (2023).
[72] M. Deutsch, Proc. Am. Acad. Art. Sci. **82**, 331(1953).
[73] V.W. Hughes, D.W. McColm, K. Ziock, R. Prepost, Phys. Rev. Lett. **5**, 63 (1960).
[74] D.B. Cassidy, A.P. Mills, Nat.(Lond.) **449**, 195 (2007).
[75] P.J. Fox, S. Jindariani, V. Shiltsev, arXiv:2203.07144 (2022).
[76] V.W. Hughes, B. Maglic, Bull. Am. Phys. Soc. **16**, 65 (1971).
[77] F. Ginzburg, U.D. Jentschura,et al., Phys. Rev. C **58** 3565 (1998).
[78] P.A. Krachkov, and A.I. Milstein, Nucl. Phys. A **971**, 71 (2018).
[79] C. Azevedo, V. P. Gonçalves, and B. D. Moreira, Phys. Rev. C **101**, 024914 (2020).
[80] R. Francener, V. P. Gonçalves, and B. D. Moreira, Eur. Phys. J. A **58**, 35 (2022).
[81] C. A. Bertulani, D. Bhandari, F.S. Navarra, Eur. Phys. J. A **60**, 43 (2024).
[82] David d'Enterria and Hua-Sheng Shao, Phys. Lett. B **842**, 137960 (2023).
[83] D. d'Enterria and G.G. da Silveira, Phys. Rev. Lett. **111**, 080405 (2013).
[84] M. Kłusek-Gawenda, P. Lebiedowicz, and A. Szczurek, Phys. Rev. C **93**, 044907 (2016).
[85] P. Lebiedowicz and A. Szczurek, Phys. Lett. B **772**, 330 (2017).
[86] M. Klusek-Gawenda, R. McNulty,, R. Schicker and A. Szczurek, Phys. Rev. D **99**, 093013 (2019).
[87] P. Jucha, M. Klusek-Gawenda and A. Szczurek, Phys. Rev. D **109**, 014004 (2024).
[88] G. Aaboud, et al., The ATLAS collaboration, Nature Physics **13**, 852 (2017).
[89] V. Gonçalves and W. Sauter Phys. Lett. B **811**, 135981 (2020).
[90] V.P. Gonçalves, D.E. Martins and M. Rangel, Eur. Phys. J. C **81**, 522 (2021).
[91] L. Schoeffel, et al., Nucl. Phys. **1005**, 121840 (2021).
[92] F.E. Low F E Phys. Rev. **120**, 582 (1960).
[93] C.A. Bertulani, Phys. Rev. C **79** 047901 (2009).
[94] E. Papageorgiu, Phys. Rev. D **40**, 92 (1989).
[95] M. Drees, J. Ellis, D. Zeppenfeld, Phys. Lett. B **223**, 454 (1989).
[96] David d'Enterria and Jean-Philippe Lansberg, Phys. Rev. D **81**, 014004 (2010).
[97] David d'Enterria, Daniel E. Martins, and Patricia Rebello Teles, Phys. Rev. D **101**, 033009 (2020).
[98] ATLAS Collaboration, Phys. Lett. B **716**, 1 (2012).

[99] CMS Collaboration, Phys. Lett. B **716**, 30 (2012).
[100] W-M Yao, et al., J. Phys. G **33**, 1 (2006).
[101] Spencer R. Klein and Joakim Nystrand, Phys. Rev. C **60**, 014903 (1999).
[102] Spencer R. Klein and Joakim Nystrand, Phys. Rev. Lett. **84**, 2330 (2000).
[103] Anthony J. Baltz, Spencer R. Klein, and Joakim Nystrand, Phys. Rev. Lett. **89**, 012301 (2002).
[104] C. Adler et al. (STAR Collaboration), Phys. Rev. Lett. **89**, 272302 (2002).
[105] B. I. Abelev et al. (STAR Collaboration), Phys. Rev. C **77**, 034910 (2008).
[106] PHENIX Collaboration, Phys. Lett. B **679**, 321 (2009).
[107] V.P. Goncalves and C.A. Bertulani, Phys. Rev. C **65**, 054905 (2002).
[108] L. L. Frankfurt, W. Koepf and M. Strikman, Phys. Rev. D **57**, (1998) 512
[109] A. Adeluyi and C.A. Bertulani, Phys. Rev. C **84**, 024916 (2011); Phys. Rev. C **85**, 044904 (2012).
[110] B. Abelev, et al. (ALICE Collaboration), Phys. Lett. B **718**, 1273 (2013).
[111] J. D. Tapia Takaki; ALICE Collaboration, AIP Conf. Proc. **1523**, 221 (2013).
[112] A. Abelev, et al., (ALICE Collaboration) Eur. Phys. J. C **73**, 2617 (2013).
[113] D. De Gruttola D (ALICE Collaboration), Nucl. Phys. A **926**, 136 (2014).
[114] V. Guzey, E. Kryshen, M. Strikman, M. Zhalov, Phys. Lett. B **726**, 290 (2013).
[115] V. Guzey, M. Strikman, and M. Zhalov, Phys. Rev. C **95**, 025204 (2017).
[116] V.P. Gonçalves, F.S. Navarra, and D. Spiering, Phys. Lett. B **791**, 299 (2019).
[117] V. P. Gonçalves, D. E. Martins, and C. R. Sena, Eur. Phys. J. A **57**, 82 (2021).
[118] Yuri V. Kovchegov, Huachen Sun, Zhoudunming Tu, [hep-ph] arXiv:2311.12208 (2023).
[119] S. Ahern, J.W. Norbury and W. Poyser, Phys. Rev. D **62**, 116001 (2000).

Fusion cross sections for the $^{7}Li + ^{27}Al$ system: fusion at energies around the Coulomb barrier

L. F. Gonzalez[1,2], J. C. Morales-Rivera[1], E. Martinez-Quiroz[2*], E. F. Aguilera[2], and P. Amador-Valenzuela[2]

[1]Facultad de Ciencias, Universidad Autónoma del Estado de México, Instituto Literario No. 100, C.P. 50000, Toluca, México
[2]Departamento de Aceleradores y Estudio de Materiales, Instituto Nacional de Investigaciones Nucleares, A. P. 18-1027, C. P. 11801, México

Abstract. In this study, we report the fusion excitation function for the $^{7}Li + ^{27}Al$ system, measured at beam energies around the Coulomb barrier (from 8.0 MeV to 11 MeV). The experiment was performed with the EN Tandem Van de Graaff Accelerator facility at ININ, Mexico. Through γ-spectra analysis, the γ-rays produced in the reaction were identified and then the corresponding fusion excitation functions of the produced nuclei were determined. Fusion-evaporation calculations were performed using the PACE2, LILITA, and CASCADE codes and compared with experimental data. The obtained total fusion cross section shows a good agreement with previously reported data. Furthermore, the higher production than predicted by fusion-evaporation codes, observed for ^{26}Mg and ^{28}Si nuclei, suggests a contribution of proton-transfer process in the production of these nuclei.

1. Introduction

In recent years, the Heavy Ion Group of the National Institute of Nuclear Research (ININ), has expanded its research to include the study of nuclear reactions using stable beams on different targets with the EN Tandem Van de Graaff Accelerator facility at ININ, in Mexico.

The main goal of these studies is to compare and complement the information about nuclear structure of nuclear radioactive beams, such as ^{7}Be and ^{8}B. For this, the elastic scattering angular distributions of the $^{9}Be + ^{51}V$ and $^{7}Li + ^{58}Ni$ systems have already been measured at energies close to their respective Coulomb barriers [1, 2].

In the framework of these studies, recently we measured the fusion cross sections for the $^{9}Be + ^{51}V$ and $^{7}Li + ^{51}V$ systems at energies around the Coulomb barrier, by using the γ-ray technique [3, 4]. The fusion cross sections, reported in Ref. [4], show a very high production of ^{52}Cr with respect to fusion-evaporation predictions of PACE2 and LILITA codes [5, 6]. These results suggest a significant contribution of other reaction channels, beyond fusion, in the production of this nucleus. Later analysis suggests, that most probably, these contributions come from the p-transfer and incomplete fusion channels.

In view of these results, we decided to study the fusion for the $^{7}Li + ^{27}Al$ system, to investigate the influence of p-transfer and breakup processes on the fusion for this system

* Corresponding author: enrique.martinez@inin.gob.mx

with the same projectile, at energies around the Coulomb barrier. Actually, no measurements have been reported previously for this system in this range of energies. Additionally, in our analysis we incorporated the calculations with the fusion evaporation code CASCADE [7].

2. Experimental Arrangement

The experiment was performed with a ^{7}Li beam produced with the 6 MV EN Tandem Van de Graaff Accelerator at ININ, in Mexico and the target was a natural composition, 6 μm foil of ^{27}Al. A diagram of the experimental setup, similar to the one used in Refs. [3, 4], is shown in figure 1, a Silicon Surface Barrier detector, SSB, placed at 150 deg, was used to measure the back-scattered charged particles, while a Hyperpure Germanium detector, Ge(HP), placed at 135 deg, was used to detect the γ-rays produced during the reaction. The beam energies used in the present experiment were 8.0, 9.0, 10.0 and 11.0 MeV.

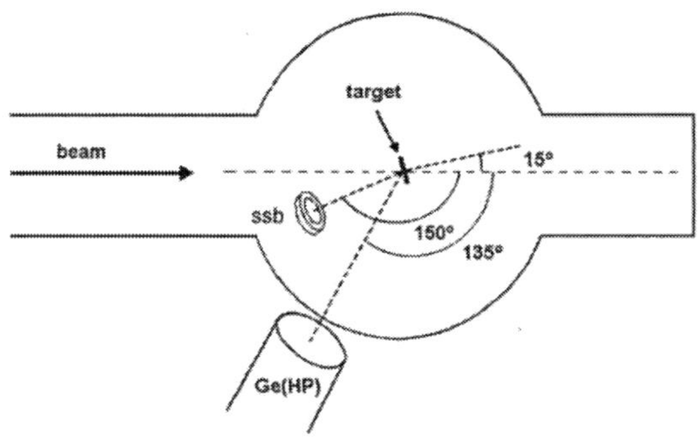

Figure 1. Diagram of the experimental arrangement used in the present experiment.

3. Results and Discussion

Figure 2 shows the measured gamma-ray spectra, for comparison, also a measured background spectrum is included.

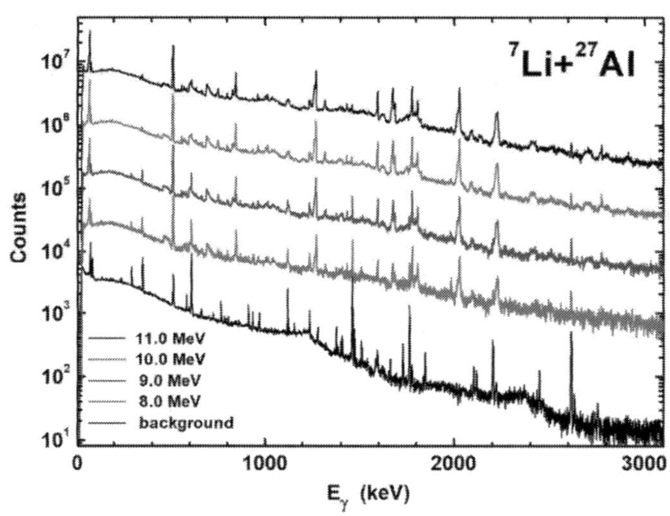

Figure 2. (color online) Measured spectra at indicated energies. A background spectrum is included for comparison.

Taking into account only the gamma-rays emitted by decays to ground state, the fusion cross sections for each gamma-ray were determined by the following formula:

$$\sigma_{fus} = \frac{N_{det}}{N_p\, N_t\, \varepsilon_{abs}} \tag{1}$$

where N_{det} is the number of gamma-rays detected (for a given energy), N_p is the number of incident projectiles, N_t is the number of atoms in the target and ε_{abs} is the absolute efficiency of the detection system.

The determined experimental excitation functions for each residual nucleus are presented in figures 3a and 3b. In both figures, the curve labeled SUM is the same, and is the sum over all residual nuclei. The experimental data (symbols) are compared with predictions of the fusion-evaporation code CASCADE (solid lines).

From comparison with experimental data, it is clear that CASCADE predictions agree very well for ^{29}Si, ^{31}P, ^{32}P and ^{32}S whose summed yield represents about 80% of the total fusion cross section at each energy. The remaining 20%, formed by ^{26}Mg, ^{29}Al, ^{28}Si and ^{33}P nuclei is underpredicted by the code, particularly for the cases of the ^{26}Mg and ^{28}Si nuclei.

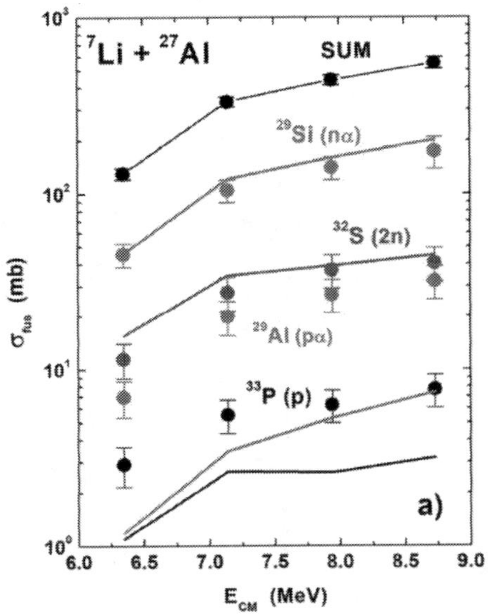

Figure 3a. (color online) Experimental fusion excitation functions for various evaporation residues (symbols), and respective CASCADE predictions (solid lines).

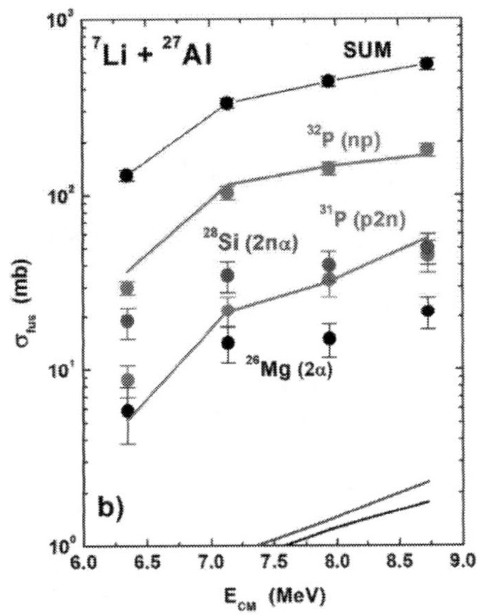

Figure 3b. (color online) Same as Figure 3a, but for different evaporation residues.

It is important to mention that predictions of the fusion-evaporation codes PACE2 and LILITA (not presented in this work), look very similar for almost all residual nuclei, with the exception of the ^{31}P nucleus. In this particular case, CASCADE reproduces very well the fusion excitation function, while PACE2 and LILITA predict very low production (fractions of mb), of this nucleus. We are looking for details of each model that could be explain this significant discrepancy.

On the other hand, all three codes, consistently underestimate the production of ^{26}Mg ^{29}Al, ^{28}Si and ^{33}P, this fact suggests that in the production of this nuclei, there could be other processes like p-transfer, incomplete fusion or other processes not yet considered. The next step of our study is to use more sophisticated models like CRC (Coupled Reaction Channels) [8], to investigate the possible contribution of the p-transfer process, that could be explain the production of ^{26}Mg and ^{28}Si, as well as to perform fusion-evaporation calculations to estimate the contribution of incomplete fusion through projectile breakup.

The total fusion cross sections are shown in figure 4 and compared with reported data from Refs. [9, 10, 11], the BPM calculations (using the formula proposed by C. Y. Wong [12]), are included for reference. It is important to mention that, during our analysis, J. Kühtreiber, *et al.*, [11], published data that cover the same interval of bombarding energies; therefore, we are including the data presented in Ref. [11]. However, their total fusion cross sections and ours are in good agreement, although there are important differences because J. Kühtreiber, *et al.* didn't report ^{28}Si as residual nucleus.

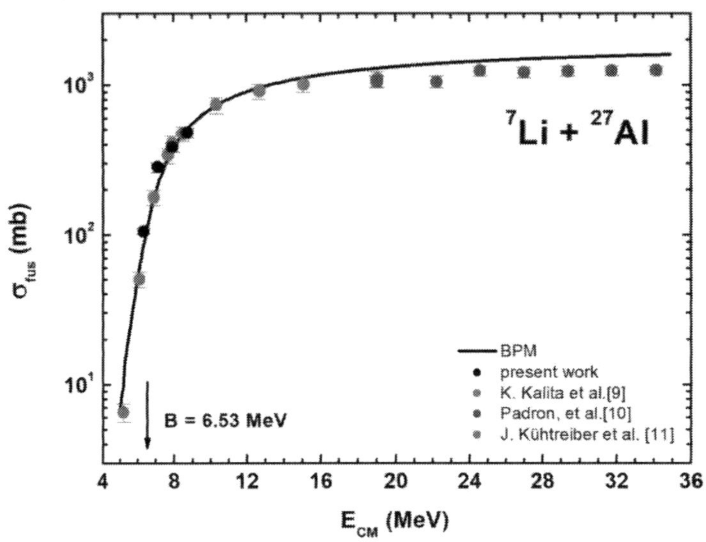

Figure 4. (color online) Experimental total fusion cross section and comparison with previously reported data, taken from Refs. [9, 10, 11] and BPM calculations as reference.

From figure 4 it is clear that our data are in good agreement with previously reported data, thus verifying the reliability of the results.

4. Summary

We measured the fusion reaction at several energies around the Coulomb barrier by using the gamma-ray technique, the studied system was ^{7}Li + ^{27}Al, the experiment was performed by using the 6 MV EN Tandem Van de Graaff Accelerator at ININ, in Mexico. The aim of this experiment was to look for evidence of over production of nuclei that could suggest a contribution of p-transfer process. The excitation functions for ^{26}Mg and ^{28}Si suggest that other processes than fusion contribute in the production of these nuclei and p-transfer channel, pickup and stripping, respectively, are good candidates. Surprisingly, we found a significant production of the residual nucleus ^{31}P (about 10% of the total fusion cross section), which is predicted with very low cross section by the PACE2 and LILITA codes.

The next step of this study is to perform calculations to determine the p-transfer and incomplete fusion contribution in the production of this and other nuclei, ^{26}Mg, ^{29}Al, ^{28}Si and ^{33}P that were not reproduced by our fusion-evaporation codes.

5. References

1. J. C. Morales-Rivera, E. Martinez-Quiroz, T. L. Belyaeva, E. F. Aguilera, D. Lizcano, P. Amador-Valenzuela, Elastic scattering of ^{9}Be+^{51}V near the Coulomb barrier, Eur. Phys. J. Web Conf. **117**, 07027 (2016). https://doi.org/10.1051/epjconf/201611707027
2. P. Amador-Valenzuela, E. F. Aguilera, E. Martinez-Quiroz, D. Lizcano, J. C. Morales-Rivera, Measurements of angular distributions for ^{7}Li elastically scattered from ^{58}Ni at energies around the Coulomb barrier, J. Phys. Conf. Ser. **876**, 012002 (2017). https://iopscience.iop.org/article/10.1088/1742-6596/876/1/012002

3.] J. C. Morales-Rivera, E. Martinez-Quiroz, E. F. Aguilera, T. L. Belyaeva, P. Amador-Valenzuela, Estudio Teórico de los canales de evaporación en la fusión del sistema ^{9}Be+^{51}V, in *Proceedings of the XXVI Congreso Técnico-Científico ININ-SUTIN*, ININ. México, December 5-7 (2016), pp 109.

4. J. C. Morales-Rivera, E. Martinez-Quiroz, P. Amador-Valenzuela, E. F. Aguilera, T. L. Belyaeva, D. Lizcano, Fusion measurements for the ^{7}Li + ^{51}V system at energies around the Coulomb barrier, J. Phys. Conf. Ser. **1078**, 012014 (2018). https://iopscience.iop.org/article/10.1088/1742-6596/1078/1/012014
5. A. Gavron, Statistical model calculations in heavy ion reactions, Phys. Rev. C **21**, pp. 230 (1980). https://doi.org/10.1103/PhysRevC.21.230
6. J. Gomez del Campo and R. G. Stokstad R G 1981 Description and use of the Monte Carlo code LILITA, ORNLTM-7295, Oak Ridge, Tennessee, USA.
7. F. Pühlhofer, On the interpretation of evaporation residue mass distributions in heavy-ion induced fusion reactions, Nucl. Phys. A **280**, pp. 267 (1977). https://doi.org/10.1016/0375-9474(77)90308-6
8. I. J. Thompson, Coupled reaction channels calculations in nuclear physics. Comp. Phys. Rep. **7**, 167, (1988). https://doi.org/10.1016/0167-7977(88)90005-6
9. K. Kalita, *et al.*, Elastic scattering and fusion cross sections for ^{7}Be,^{7}Li +^{27}Al systems, Phys. Rev. C **73**, 024609 (2006). https://doi.org/10.1103/PhysRevC.73.024609
10. I. Padron, *et al.*, Fusion of stable weakly bound nuclei with ^{27}Al and ^{64}Zn, Phys. Rev. C **66**, 044608 (2002). https://doi.org/10.1103/PhysRevC.66.044608
11. J. Kühtreiber, *et al.*, ^{6,7}Li+^{27}Al reactions close to and below the Coulomb barrier, Phys. Rev. C **103**, 064605 (2021). https://doi.org/10.1103/PhysRevC.103.064605
12. C. Y. Wong, Interaction Barrier in Charged-Particle Nuclear Reactions, Phys. Rev. Lett. **31**, pp 766 (1973). https://doi.org/10.1103/PhysRevLett.31.766

Characterization of thin ^{nat}Si targets

J. R. Fuentes-Carreón[1], *J.* Mas-Ruiz[1,2,*], *D. J.* Marín-Lámbarri[1], *Arcadio* Huerta[1], *Stephen* Muhl[3], *A.* Rodríguez-Gómez[1], and *Efraín* Chávez[1]

[1]Instituto de Física (UNAM), Sendero Bicipuma, C.U., Coyoacán, 04510 Ciudad de México, CDMX
[2]Instituto de Ciencias Nucleares (UNAM), Cto. Exterior, C.U., Coyoacán, 04510 Ciudad de México, CDMX
[3]Instituto de Investigaciones en Materiales (UNAM), Circuito de la, Circuito Exterior, Investigación Científica S/N, C.U., 04510 Ciudad de México, CDMX

Abstract. The production of ^{26}Al is important in the field of stellar nucleosynthesis. The study for the production of ^{26}Al from the ^{28}Si(d, α) reaction at energies that are close to the Coulomb barrier requires thin silicon targets. This work reports the characterization of thin silicon targets by using the RBS (Rutherford Backscattering Spectrometry) technique, with proton beams, ^{12}C and ^{28}Si, taking into account a scattering angle of 140° at a range of energies ranging from 841 keV to 7105 keV. This experiment was carried out at the Nuclear Physics line of the National Accelerator Mass Spectrometry Laboratory (LEMA) of the Institute of Physics of UNAM. The first three targets to be characterized, labelled C11, C21 and C31 were fabricated at the Instituto de Investigaciones en Materiales (IIM-UNAM). The fourth target F11 was manufactured at the Microscopy Laboratory of the Institute of Physics (IF-UNAM). The thickness of the targets C11, C21 and C31 was 10^{16} atoms/cm^2. The thickness of the F11 target is 10^{17} atoms/cm^2, therefore, this target is suitable to be used in the production of ^{26}Al.

1 Introduction

Rutherford back scattering spectrometry (RBS) is an important nuclear-based technique used in ion beam analysis (IBA) in a wide range of studies, such as angular distributions and cross sections [1]. Among its main applications is the characterization of thin films, which allows us to obtain accurate measurements of the composition and amount of elements in a sample [2–5]. It can also analyze the distribution of the elements and how deep they are in the sample over a range of nanometers to micrometers.

In the present work, thin silicon targets are characterized with the motivation of being used for the production of ^{26}Al through the reaction ^{28}Si(d,α), it is necessary to have a thin silicon film of the order of 10^{17} monolayers for irradiation to be suitable. The results obtained from the characterization of four silicon targets are presented, as well as the measurement and analysis process.

*e-mail: javier.masruiz91@gmail.com

2 Materials and methods

The experiment was carried out at the Accelerator Mass Spectrometry Laboratory(LEMA) [6] of the Institute of Physics of the National Autonomous University of Mexico, using the line dedicated to nuclear physics and scattering experiments. Beams of different energies can be obtained in the accelerator depending on the state of charge selected at the exit of the accelerator tank [7]; the details of the acceleration process can be consulted in [8].

The final kinetic energy E_c of the incident beam on the target inside the reaction chamber is given by equation 1:

$$E_h(keV) = E_e + (1 + q)V_t \tag{1}$$

where E_e = 35 keV is the kinetic energy of beam extraction from the ion source, q is the charge state acquired by the particles after losing electrons inside the accelerator tank, and V_t (kV, max 1 MV) is the accelerator terminal voltage.

The three beams used in this experiment were produced at LEMA. A titanium hydride (TiH_2) cathode was used for the proton beam, and graphite and silicon cathodes were used for the ^{12}C and ^{28}Si beams, respectively.

Four targets were used for the characterization: three targets labeled (C11, C12, and C31), manufactured at the Materials Research Institute (IIM-UNAM), and another target labeled (F11) at the Microscopy Laboratory of the Physics Institute (IF-UNAM).

The targets are composed of graphite sheets as a substrate, and a thin film of silicon was deposited on them. A thin film of gold was also deposited on them for normalization purposes. An *Ortec ULTRA SERIES C* silicon detector was placed at 140°. A typical ORTEC electronic chain was used for the measurements, coupled to a UCS30 ADC.

3 Results and Discussion

Seventeen measurements were taken at LEMA for three targets (C11, C21, C31) and five measurements for one target (F11). For the acquisition of the spectra, the USX 1.2.00 spectral techniques software was used. The spectra were saved on a USB for their respective analysis, naming and numbering each one according to the beam used (protons, ^{12}C, ^{28}Si), the beam energy, and the date of acquisition of that spectrum for each target.

3.1 Target layers characterization

This section presents the graphs corresponding to the analyses conducted on targets C11, C21, C31, and F11, for the various beams and energies used.

3.1.1 Target C11 characterization

841 keV and 865 keV proton beams: For the characterization of the C11 target, proton beams with energies of 841 keV and 865 keV were used with a scattering angle of 140°. The simulated spectrum was fitted to the experimental data. For the silicon peak, the thickness is 16 monolayers, corresponding to a density value of 746.192 ng/cm^2, assuming that 1 monolayer is equivalent to 10^{15} atom/cm^2. For the gold peak, the thickness is 5 monolayers, with a density value of 1635.355 ng/cm^2. The simulated spectrum, fitted to the experimental data for the 841 keV proton energy, is shown in Figure 1 (left side).
5085 keV ^{12}C beam: Figure 1 (right side) shows the simulation and spectrum acquired for target characterization using the ^{12}C beam. The thickness values for the silicon and gold layers are similar to those obtained with the proton beam.

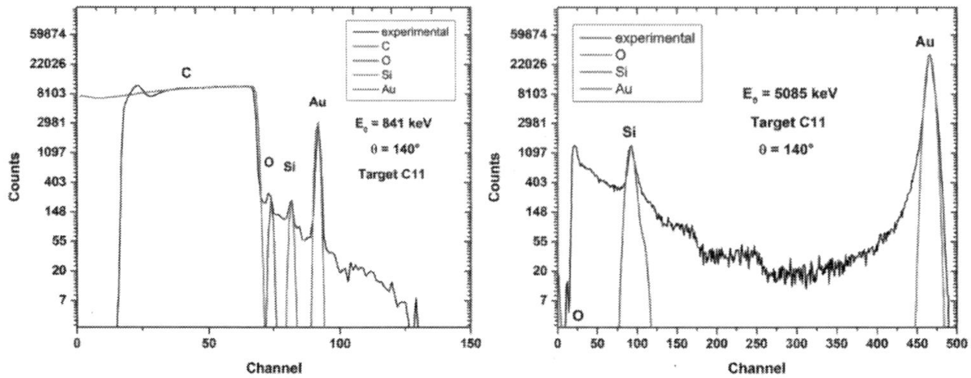

Figure 1. Experimental (black line) and simulated (color line) RBS spectra of a proton beam at an energy of 841 keV (left) and 5085 keV of ^{12}C beam (right) measured at 140° for the C11 target.

3.1.2 Target C21 characterization

1035 keV proton beam: For the characterization of the C21 target, with a proton beam at an energy of 1035 keV, the simulated spectrum was fitted to the experimental data. For the silicon peak, the thickness is 65 monolayers, corresponding to a density value of 3031.405 ng/cm^2. For the gold peak, the thickness is 12 monolayers, with a density of 3924.852 ng/cm^2. The simulated spectrum, fitted to the experimental data, is shown in figure 2.

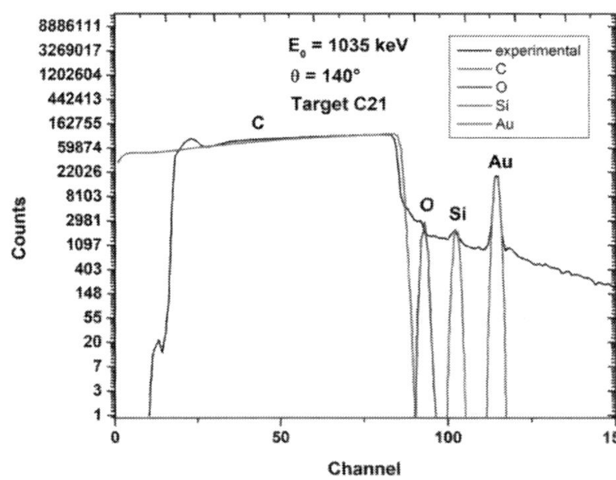

Figure 2. Experimental (black line) and simulated (color line) RBS spectra of a proton beam at an energy of 1035 keV, measured at 140° for the target C21.

3.1.3 Target C31 characterization

885 keV and 901 keV proton beam: For the characterization of the C31 target, with a proton beam at an energy of 885 keV, is shown in figure 3 (left side), and at an energy of 901 keV (right side). The simulated spectrum was fitted to the experimental data. For the silicon peak, the thickness is 43 monolayers, corresponding to a density value of 2005.391 ng/cm^2. For the gold peak, the thickness is 9 monolayers, with a density of 2943.639 ng/cm^2. The simulated spectrum, fitted to the experimental data, is shown in Figure 3.

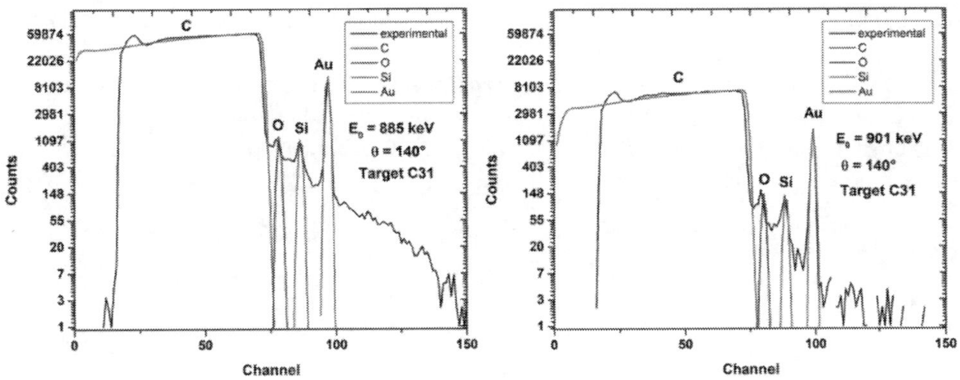

Figure 3. Experimental (black line) and simulated (color line) RBS spectra of a proton beam at energies of 885 keV and 901 keV, measured at 140° for the target C31.

The C11, C21, and C31 targets have a thickness of 10^{16} atoms/cm^2 according to the thickness values obtained from the number of monolayers for each peak corresponding to the elements that compose these targets, manufactured at IIM-UNAM.

3.1.4 Target F11 characterization

1000 keV proton beam: The characterization of target F11 using a 1000 keV proton beam is shown in figure 4 (left side), with a scattering angle of 140°. The simulated spectrum was fitted to the experimental data. For the silicon peak, the thickness is 125 monolayers, corresponding to a density value of 5829.625 ng/cm². For the gold peak, the thickness is 1.6 monolayers, with a density of 523.314 ng/cm².

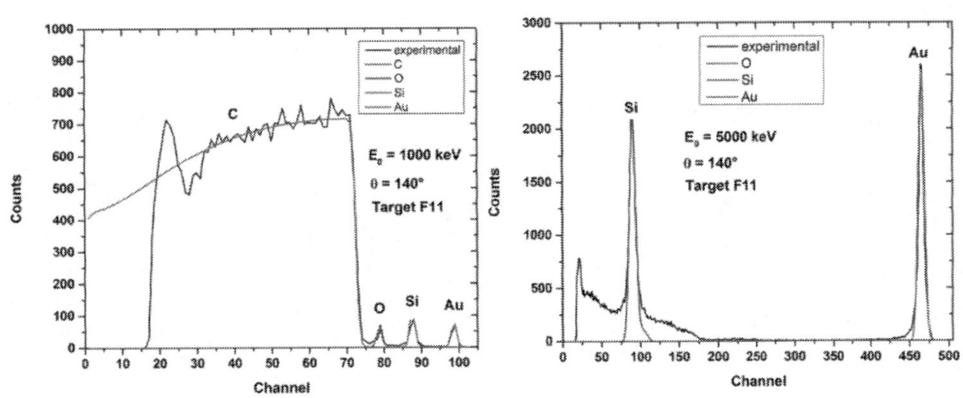

Figure 4. Experimental (black line) and simulated (color line) RBS spectra of a proton beam (left) at an energy of 1000 keV and ^{12}C beam (right) at an energy of 5000 keV, measured at 140° for the F11 target.

5000 keV ^{12}C beam: For the F11 target, a ^{12}C beam at an energy of 5000 keV was used. The simulated spectrum was fitted to the experimental data, as shown in figure 4 (right side).

For the silicon peak, the thickness is 112 monolayers, corresponding to a density value of 5223.344 ng/cm². The gold peak has the same thickness.

The target F11 has a thickness of 10^{17} atoms/cm^2 according to the thickness values obtained from the number of monolayers for each peak corresponding to the elements that make up the target, manufactured at IF-UNAM.

The table 1 shows the summary of targets and beams used in this work, with the thickness values in monolayers. The target F11 was the only one with a thickness of around 10^{17} atoms/cm^2, as the irradiation demands. So, all the manufacturing parameters were saved to replicate more targets like this one.

Table 1. Table of thickness values of the layers and beams used for characterization.

Beam	**Protons**		**^{12}C**		**^{28}Si**
Layer	Gold	Silicon	Gold	Silicon	Gold
Target C11	5 ML	16 ML	5 ML	16 ML	*
Target C21	12 ML	65 ML	*	*	12 ML
Target C31	9 ML	43 ML	*	*	*
Target F11	1.6 ML	125 ML	1.6 ML	112 ML	*

* Not measured data; ML-monolayers

4 Conclusions

Four targets [C11, C21, C31 (IIM-UNAM)] and [F11 (IF-UNAM)] were successfully characterized with three different beams, (protons, ^{12}C and ^{28}Si) at a scattering angle of 140°, the energy beam ranges from 841 keV to 7105 keV. A good simulation was performed using the "SIMNRA" software. The characterization with a proton beam revealed that the targets C11, C21, and C31 had silicon thicknesses of (16, 65, 43) $\times 10^{15}$ atoms/cm^2 and gold thicknesses of (5, 12, 9) $\times 10^{15}$ atoms/cm^2, respectively. We measured for target F11 a silicon thickness of 125×10^{15} atoms/cm^2 and gold 1.6×10^{15} atoms/cm^2. We obtain the same thickness in the gold and silicon layer for targets C11 and C21, using different beam species. For the measure with ^{12}C in the target C11 we obtain a silicon thickness of 16×10^{15} atoms/cm^2 and gold thickness of 5×10^{15} atoms/cm^2. The analysis of target F11 revealed thicknesses of 112×10^{15} atoms/cm^2 and 1.6×10^{15} atoms/cm^2 of silicon and gold, respectively. The thickness value of the gold layer for all targets was as expected; however, for the silicon layer, only target F11 had an average density of 1.19×10^{17} atoms/cm^2. This target can be used for the production of ^{26}Al.

Acknowledgments

This work has been possible thanks to the financial support from UNAM projects number IG102023, IG101423, IN112023, and TA100524. This work was partially financially supported by CONACYT through contracts: ApoyosLNC-2023-58, CBF2023-2024-862, and PIIF-5-24 by IFUNAM.

References

[1] J. Mas-Ruiz, K.G. Zayas-Bazán, P.G. Zayas-Bazán, A. Huerta, J. Sastré-Hernández, D.J. Marín-Lámbarri, L. Acosta, E. Andrade, C. Solís, E.R. Chávez Lomelí, Probing the elastic scattering differential cross section for al+ p at backward angles in a low energy regime, Universe **9**, 438 (2023).

[2] G. Umapathy, S. Ojha, K. Rani, M. Thakur, R. Mahajan, N. Chopra, D. Kanjilal, Composition profile of thin film target by Rutherford backscattering Spectrometry, in *Proceedings of the 61st DAE-BRNS Symposium on Nuclear Physics, Kolkata, India* (2016), pp. 5–9

[3] P. Lorenzen, H. Rothard, K. Kroneberger, J. Kemmler, M. Burkhard, K. Groeneveld, Target characterization by fast ion impact, Nuclear Instruments and Methods in Physics Research Section A: Accelerators, Spectrometers, Detectors and Associated Equipment **282**, 213 (1989).

[4] M. Döbeli, A. Lombao, D. Vetterli, M. Suter, Thin layer analysis with heavy ion rbs, Nuclear Instruments and Methods in Physics Research Section B: Beam Interactions with Materials and Atoms **89**, 174 (1994).

[5] A.S. Gorzalski, C. Donley, O. Coronell, Elemental composition of membrane foulant layers using eds, xps, and rbs, Journal of Membrane Science **522**, 31 (2017).

[6] C. Solís, E. Chávez-Lomelí, M. Ortiz, A. Huerta, E. Andrade, E. Barrios, A new AMS facility in Mexico, Nuclear Instruments and Methods in Physics Research Section B: Beam Interactions with Materials and Atoms **331**, 233 (2014).

[7] R. Middleton, A negative ion cookbook, University of Pennsylvania, unpublished (1989).

[8] M. Ortiz, C. Solís, E. Chávez, Espectrometría de Masas con Aceleradores (Laboratorio Nacional, Universidad Nacional Autónoma de México, 2018)

EPJ Web of Conferences **333**, 01004 (2025) https://doi.org/10.1051/epjconf/202533301004
XLVI Symposium on Nuclear Physics 2025

A Semimicroscopic Algebraic Cluster Model for heavy nuclei applied to ^{162}Dy

P.O. Hess[1]

[1]Frankfurt Institute for Advanced Studies, J. W. von Goethe University, Hessen, Germany

Abstract. The main objective of the present contribution is to describe ^{162}Dy within the *Semimicroscopic Algebraic Cluster Model* for heavy nuclei (SACM-h), intentionally using an simple Hamiltonian in an $SU(3)$ limit in order to provide a clear insigt into the basic structure of ^{162}Dy. It is shown that an important ingredient is not only the Hamiltonian but more importantly the Hilbert space, taken as the shell model space. For heavy nuclei the pseudo-$SU(3)$ model ($\widetilde{SU}(3)$) is applied. The spectrum and $B(E2)$-transitions are reproduced. Of particular interest is the 0^+ spectrum. It is shown that by knowing the content of the $\widetilde{SU}(3)$, the 0^+ spectrum is understood including its multiplicities. In particular, the 0^+ spectrum is described fairly well up to the clustering of 0^+ states.

1 Introduction

The main objective of is contribution is to apply the *Semimicroscopic Algebraic Cluster Model* for heavy nuclei (SACM-h) [1] to ^{162}Dy. The model uses a microscopic shell-model space, but a phenomenological Hamiltonian. Intentionally we will use a simple Hamiltonian in an $SU(3)$ limit in order to highlight the basic structural aspects and ingredients for describing heavy nuclei in this model, especially the effects of taking into account the *Pauli Exclusion principle* (PEP). The space is constructed, using the well known *pseudo-*$SU(3)$ model (denoted from now on as $\widetilde{SU}(3)$) [2, 3]. Instead of $\widetilde{SU}(3)$ one probably can use the proxy-$SU(3)$ [4, 5], which also takes into account the intruder orbitals. It will be explained how the microscopic model space is obtained. As a particular application, ^{162}Dy is considered, which has been the center of frequent theoretical and experimental studies [6–9], but is not restricted to this nucleus. One particular feature of this nucleus is the accumulation of 0^+ states at low energy, which is drawing much recent attention (see some contributions in this conference).

The structure of the Hamiltonian will be kept simple, because our main intention is not to reproduce in every detail the structure of ^{162}Dy, but rather to provide a better unobstructed understanding of the spectrum and $B(E2)$ transitions of this nucleus. Some remarks to the IBA model [10] will be presented, stating that it is indistinguishable to the geometric model with a finite cutoff of the boson number, as demonstrated in [11], therefore, the IBA works quite well. The mathematical relation of the IBA and the geometric model of the nucleus was shown already in [12]. However, for well deformed mid-shell nuclei it cannot be stated that the IBA can be related to a microscopic model, because it neglects the exchange effects (PEP) due to not taking into account the substructure of the bosons in terms of two fermions.

It will be stressed, that the PEP in a many fermion system is of utmost importance, as in a nucleus. For example, in [13] the wave function of states in the ground state band were analyzed, taking into account or not the PEP, resulting in a clear picture that not taking into account the PEP the intrinsic states of the band are not consistently described. For more discussion, please consult [14–17].

This contribution is organized as follows: in Section 2 the microscopic space will be constructed. The $\widetilde{SU}(3)$ will be used (analogous to light nuclei). In Section 3 The Hamiltonian and $B(E2)$-transition operator will be defined, also the microscopic model space is constructed. The SACM-h will be applied to ^{162}Dy and compared to experimental data. Only $0\hbar\omega$ states and $B(E2)$-values are calculated. Energies and $B(E2)$ transitions will be determined and compared to their structure within the IBA. In Section 4 Conclusions will be drawn.

2 The microscopic model space

To simplify the consideration, only the valance shell will be considered, i.e., no $n\hbar\omega$ excitations, with $n > 0$, and no negative parity states which would require the $1\hbar\omega$ shell.

Because ^{162}Dy is a heavy nucleus, we use the $\widetilde{SU}(3)$ model [2, 3], following closely [1]. In a nutshell, the orbitals in each harmonic oscillator shell of the shell model are separated into normal states and intruders. The nucleons in the intruder orbitals are occupied by nucleon pairs, which are difficult to break due to the large j-spin. Due to this property they are treated as spectators, as reasoned in detail in [1–3]. The normal orbitals are renamed, introducing a pseudo angular momentum $\widetilde{l}$ and pseudo-spin $\widetilde{s}$. Doing so, one notices that the content of the remaining normal orbitals of the shell η are identical to the content of a harmonic oscillator shell $\widetilde{\eta} = \eta - 1$, hence the name pseudo-$SU(3)$. In this mapping, the spin-orbit interaction $\kappa(l \cdot s)$, which is large for heavy nuclei, maps to a pseudo-spin-orbit interaction $\widetilde{\kappa}(\widetilde{l} \cdot \widetilde{s})$ [18], with $\widetilde{\kappa} << \kappa$, which explains the justification for the use of pseudo-$SU(3)$ symmetry. The sum of the intruder orbitals plus the normal orbitals in each shell $\widetilde{\eta}$ is still the complete shell model space, nothing is put under the rug. It is just a mapping/renaming, with the great advantage that one can use the great power of symmetry again, as for light nuclei.

The protons and neutrons occupy different shells and, thus, are considered separately and are later combined. Assuming a valence shell with radial quantum number $\widetilde{\eta}$ ($\widetilde{\eta} = \widetilde{\eta}_p$ for protons and $\widetilde{\eta} = \widetilde{\eta}_n$ for neutrons), the relevant group chain to consider, separately for the proton and neutron sector, is

$$
\begin{array}{ccc}
U((\tilde{\eta}+1)(\tilde{\eta}+2)) & \supset & U(\frac{1}{2}(\tilde{\eta}+1)(\tilde{\eta}+2)) \otimes U_S(2) \\
\left[1^N\right] & & [\widetilde{h}] \qquad\qquad [h] \\
U(\frac{1}{2}(\tilde{\eta}+1)(\tilde{\eta}+2)) & \supset & \widetilde{SU}(3) \\
{[h]} & & (\lambda, \mu) \quad ,
\end{array}
\tag{1}
$$

where $[h]$ is a short hand notation for $[h_1, h_2]$ and the tilde denotes the conjugate Young diagram where rows and columns are interchanged. The group chain in (1) points out the relation of the spin-part (S) and the orbital part, such that the complete state is anti-symmetric ($\left[1^N\right]$, with N as the number of protons or neutrons in the pseudo-shell $\tilde{\eta}$). The lowest states in energy are the most anti-symmetric states in $U_S(2)$. In the second line, the reduction of the orbital group to the $\widetilde{SU}(3)$ group is indicated. Programs for the reductions appearing in (1) are available [19, 20].

The model space of the SACM-h is constructed by assuming a cluster structure of the nucleus and constructing an overlap to the shell model space, which contains the lowest lying $\widetilde{SU}(3)$ irreps. This method is a manner to reduce the size of the model space. Because the Hamiltonian will not contain interactions sensible to the cluster structure and the Hamiltonian will have a pure $\widetilde{SU}(3)$ structure, one can also use $\widetilde{SU(3)}$ of ^{162}Dy. The cluster structure used for ^{162}Dy is α + ^{162}Gd. In order to satisfy the *Wildermuth condition* [21], the minimum value for the relative oscillation quanta has to be determined. The cluster irreps of α and the Gd clusters are multiplied with this relative oscillator irrep and then the overlap to the shell model is taken. All these steps have to be realized separately for protons and neutrons, see [1], and the numbers are deduced below. For more details we recommend to consult [1]

In order to obtain the number of normal protons (neutrons) in a pseudo-shell, we use the Nilsson diagram and draw a vertical line, at the prolate deformation for the nucleus listed in [22]. For the Nilsson diagrams we use the figures of [23]. Then, the nucleons are filled in from below. Counting the number of normal and unique orbitals filled by the nucleons, we obtain the number of protons (neutrons) in the normal orbitals as twice the number of filled normal orbitals (two nucleons in each orbital).

What follows is a brief example on how to deduce the shell model space for ^{162}Dy: In the proton case of ^{162}Dy the pseudo shell is $\tilde{\eta} = 3$. Thus the first group chain in (1) is $U(30) \supset U(10) \otimes U_S(2)$. At the deformation $\beta_2 = 0.28$ [22], there are 10 protons in the pseudo-shell and, thus, the irrep of $U_S(2)$ is $\left[5^2\right]$ and, due to the complementarity, the irrep of $U(10)$ is $\left[2^5\right]$. The reduction of $U(10)$ to $\widetilde{SU}(3)$ is obtained using the routines in [19, 20]. As a result, the $\widetilde{SU}(3)$ irreps with the largest eigenvalue of the second Casimir operator are $(10,4)_\pi$, $(12,0)_\pi$ and $(7,7)_\pi$. Because we deal with a well deformed nucleus, the quadrupole-quadrupole interaction dominates, resulting in the largest $\widetilde{SU}(3)$ to be lowest in energy.

For the neutron part we proceed in an analogous manner. For neutrons, the pseudo-shell is $\tilde{\eta} = 4$ and the resulting number of neutrons in this pseudo-shell turns out to be also 10. The orbital group is $U(15)$ with the irrep $\left[2^5\right]$, which has to be reduced to the pseudo-$SU(3)$ group. the largest irreps turn out to be $(20,4)$, $(22,0)$, $(17,7)$.

When combining the proton part with the neutron one, we only take into account the linear coupling $(\tilde{\lambda}_\pi + \tilde{\lambda}_\nu, \tilde{\mu}_\pi + \tilde{\mu}_\nu)$, which correspond to the alignment of the proton with the neutron fluid [24] (see also [1]). This gives for the larges total $SU(3)$ irreps the list $(30,8)$, $(32,4)$, $(27,11)$ and $(34,0)$.

The final obtained model space of the valence shell is listed in Table 1. The largest irrep of the $\widetilde{SU}(3)$ group is (30,8). Note, that this irrep contains the bands K = 0,2,4,6,8, i.e., the K=2 band, the gamma band, is within the same irrep as the ground state band ($K = 2$). The $K = 0$ excited band, i.e. the β-band, belongs to the (32,4) irrep. Because the $\widetilde{SU}(3)$-quadrupole operator connects only states within the same irrep, there are transitions from the γ-band to the ground state band but not from the β-band to the ground state and γ-band. Note, that the irrep (32,4) appears with a multiplicity of 2. Also important is that the (24,14) irrep appears three times, leading to three degenerate 0^+ bands within this model. Due to the similarity of $\tilde{\lambda}$ and $\tilde{\mu}$ in the irrep, these states are not pushed up as strongly as other $\widetilde{SU}(3)$ irreps, though, with a larger C_2. The horizontal line in the table indicates more irreps.

Finally, some comments are due to the model space of the IBA. We refer to [25–27] where it is conclusively shown that the IBA has a *fatal flaw* [25–27], i.e., for mid-shell nuclei it does not take into account exchange effects: Let us denote by Ω the degeneracy of the orbital states in a shell η, given by $\Omega = \frac{1}{2}(\eta+1)(\eta+2)$, and by $k = 2,4$, additional degrees of freedom, which is 2 when only spin is taken into account (we use the notation $\eta = \tilde{\eta}$) and 4 when spin-isospin is taken into account (in the case of light nuclei). As soon as the shell is filled by a third, i.e., $2N = \frac{k\Omega}{3}$, a limit is reached. Any further addition of a particle has as the lowest

Table 1. Selected irreps of the shell model space content for the $0\hbar\omega$ space. The C_2 is the eigenvalue of the second order Casimir operator of $\widetilde{SU}(3)$. The "multip." denotes the multiplicity of the corresponding $\widetilde{SU}(3)$ irrep and the eigenvalue C_3 of the third order Casimir operator is listed in the 3rd column. The last column lists the multiplicity of the corresponding $\widetilde{SU}(3)$ irrep. The irreps (λ, μ) also indicate the deformation type of the nucleus: For $(\lambda, 0)$ the nucleus is prolate, for $(0, \mu)$ it is oblate and for (λ, λ) it is triaxial. Anything in between tends to the one or other type of deformation.

$(\tilde{\lambda}, \tilde{\mu})$	C_2	C_3	multip.
(26,10)	1144	50960	7
(14,22)	1096	-25864	1
(29,7)	1201	68816	4
(21,17)	1201	14284	2
(24,14)	1222	35750	3
(28,9)	1228	63308	2
(34,0)	1258	89318	1
(27,11)	1261	56576	1
(32,4)	1276	85484	2
(30,8)	1318	76538	1

state in energy an $SU(3)$-irrep (λ, μ) with λ and μ both different from zero. One can construct the whole shell model space, using (1) and all possible irreps $[h]$. There appear irreps of the type $(\lambda, 0)$, however, all of them are at a substantial higher energy. The contribution of J. N. Ginocchio [28] cannot change this finding. In [28] two types of pair algebras (the $Sp(6)$ and the $SO(8)$ algebra) were constructed. The $Sp(6)$ group contains the $SU(3)$ group of the shell model and has to be chosen. However, Ginocchio excluded it because it does not permit all combinations of bosons. Therefore, the $SO(8)$ group was chosen as the one describing IBA and all follow up works restrict to this group. We do not agree to this conclusion. The $SO(8)$ algebra is an interesting case but describes configurations, which according to the shell model, lie at high energy and are, thus, not permissible for the description of collective states at low energy.

Table 2. The parameters, including the deformation value [22]. $e_{\rm eff}$ is the effective charge, used in the quadrupole transition. The parameters were obtained by fitting to the lowest states. The units of χ, b, c and d are un units of MeV.

χ	b	c	β_2	d	$e_{\rm eff}$
0.063188	0.20245	0.01306	0.281	8.1651×10^{-6}	0.93

3 The Hamiltonian and comparison to experiment

In order to illustrate basic effects of the interaction terms of the model, we will not use the most general possible Hamiltonian but rather use a quite simple one, namely

$$\begin{aligned} \boldsymbol{H} &= -\frac{\chi}{2}\boldsymbol{C}_2(\widetilde{SU}(3)) + b\boldsymbol{K}^2 + c\boldsymbol{L}^2 + d\boldsymbol{C}_3(\widetilde{SU}(3)) \\ E &= -\frac{\chi}{2}C_2(\lambda,\mu) + bK^2 + cL(L+1) + dC_3(\lambda,\mu) \quad . \end{aligned} \tag{2}$$

Table 3. Comparing the theoretical energies with the experiment. The energies are given in MeV.

J^{π}_{k}	$(\widetilde{\lambda},\widetilde{\mu})$	Theory [MeV]	Exp. [MeV] (ENSDF)
0^+_1	(30,8)	0.	0.
2^+_1	(30,8)	0.078	0.081 [7]
4^+_1	(30,8)	0.261	0.266 [7]
6^+_1	(30,8)	0.549	0.548 [7]
2^+_2	(30,8)	0.888	0.880 [7]
3^+_1	(30,8)	0.967	0.963 [7]
4^+_2	(30,8)	1.071	1.061 [7]
5^+_1	(30,8)	1.202	1.183 [7]
6^+_2	(30,8)	1.358	1.324 [7]
$0^+_{2,3}$	(32,4)	1.400	1.400, 1.666 [9]
2^+_3	(32,4)	1.479	1.453 [7]
4^+_3	(32,4)	1.661	1.574 [7]
6^+_3	(32,4)	1.949	1.768 [7]
1^+_1	(27,11)	1.866	1.746 [7]
2^+_1	(27,11)	1.919	1.782 [7]
3^+_1	(27,11)	1.997	1.840 [7]
4^+_1	(27,11)	2.102	1.951 [7]
5^+_1	(27,11)	2.232	-
6^+_1	(27,11)	2.389	-
0^+_4	(34,0)	1.896	2.128 [9]
$0^+_{4,5,6}$	(24,14)	2.700	2.498, 2.656, 2.861 [9]

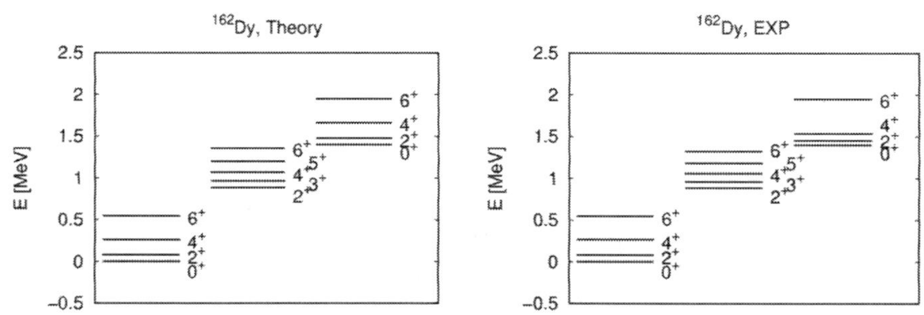

Figure 1. Theoretical spectrum (left panel) compared to experiment (right panel), for the ground state, γ and β bands.

No $n\hbar\omega$ term appears because shell excitations are not taken into account. The χ describes the strength of the quadrupole-quadrupole interaction, the b terms takes into account the K-splitting between states with different K-values in the same $\widetilde{SU}(3)$ irrep (λ,μ), the c term corresponds to the angular momentum and the last term corresponds to the third order Casimir operator. As we will see later, this term is responsible for the particular ordering of the 0^+ states in ^{162}Dy. The eigenvalue $C_2(\widetilde{\lambda},\widetilde{\mu})$ of the second order Casimir operator is $\left(\widetilde{\lambda}^2 + \widetilde{\lambda}\widetilde{\mu} + \widetilde{\mu}^2 + 3\widetilde{\lambda} + 3\widetilde{\mu}\right)$, while the eigenvalue $C_3(\widetilde{\lambda},\widetilde{\mu})$ of the third order Casimir operator is $\left(\widetilde{\lambda} - \widetilde{\mu}\right)(2\widetilde{\lambda} + \widetilde{\mu} + 3)\left(\widetilde{\lambda} + 2\widetilde{\mu} + 3\right)$. The second row in (2) is the corresponding energy eigen-

https://doi.org/10.1051/epjconf/202533301004

Table 4. $B(E2)$ transitions, theory compared to experiment. Units are in WU.

$J_i^\pi \rightarrow J_f^\pi$	Theory [WU]	Exp. [WU]
$2_1^+ \rightarrow 0_1^+$	202.56	204 ± 0.03 [29]
$4_1^+ \rightarrow 2_1^+$	287.69	289 ± 0.12 [29]
$6_1^+ \rightarrow 4_1^+$	313.52	301 ± 0.17 [29]
$2_2^+ \rightarrow 0_1^+$	9.38	4.6 ± 0.3 [29]
$2_2^+ \rightarrow 2_1^+$	14.60	8.2 ± 0.5 [29]
$2_2^+ \rightarrow 4_1^+$	0.89	0.63 ± 0.04 [29]
$3_1^+ \rightarrow 2_1^+$	0.35	< 8 [9]
$3_1^+ \rightarrow 4_1^+$	0.63	< 0.3 [9]
$4_2^+ \rightarrow 2_1^+$	4.97	$1.9^{+2.4}_{-1.2}$ [9]
$4_2^+ \rightarrow 4_1^+$	17.92	$4.4^{+3.6}_{2.8}$ [9]
$4_2^+ \rightarrow 6_1^+$	2.12	$1.4^{+1.0}_{-0.8}$ [9]
$4_2^+ \rightarrow 2_2^+$	119.45	140^{+110}_{-87} [9]
$2_3^+ \rightarrow 4_1^+$	0.	$0.79^{+0.56}_{-0.51}$ [9]

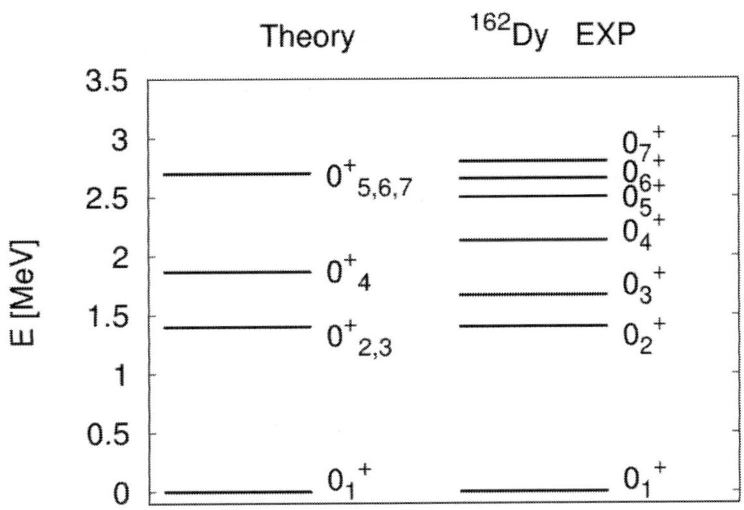

Figure 2. The first seven 0^+ states, theory compared to experiment.

value, which is used to adjust to the experimental spectrum. The parameters, obtained via a fit to experimental data, are listed in Table 2 and the corresponding comparison of theory to experiment in given in Fig. 1, while numerical values of the adjusted energies, compared to observation, are listed in Table 3. The χ and d parameters are responsible for adjusting the 0_2^+ state and the 0^+ triplet of (24,14), the a parameter to obtain the 2_1^+ state and the b parameter for the 2_2^+ state.

As seen, the adjustment is quite acceptable. Fig. 2 lists the first seven 0^+ states, again theory compared to experiment. By inspecting Table 1 we note that the (32,4) irrep is doubly degenerated, i.e., it will provide two 0^+ bands, one being associated to the β band. This degenerate state has to be compared to the first and second excited 0^+ states observed in experiment. The next theoretical calculated 0_4^+ band is single degenerated, which has to be compared to the 0_4^+ experimental states. Finally, the next (24,14) irrep is triple degenerated, which has to be compared to the last three plotted states in Fig. 2. These are closely lying states in energy, which agrees to the theoretical state, which is exactly degenerate. Because a simple Hamiltonian is used, these states (and the one of $0_{2,3}^+$) do not split. A more general Hamiltonian, which includes for example $O(6)$ interactions (mixing the $\widetilde{SU}(3)$ states) may split these states, to agree closely to experiment. Nevertheless, the basis structure of the experimental spectrum is quite well reproduced, as for example the near degeneracy of the last depicted excited 0^+ states. These states are coming down due to the third order Casimir operator, which has a factor $(\lambda - \mu) = (24 - 14) = 10$, a significant lower value than in the other irreps.

The quadrupole operator is defined as [30]:

$$\widetilde{Q}^a_m = e_{\text{eff}}\sqrt{3}\widetilde{C}^{(1,1)}_{2m} \quad , \tag{3}$$

where $\widetilde{C}^{(1,1)}_{2m}$ is a generator of $\widetilde{SU}(3)$ and e_{eff} is an effective charge. The *triple reduced matrix element* of this operator is given by [30]. The quadrupole operator in (3) is the portion of the *physical* quadrupole operator which acts within the valence shell only. One further parameter, the effective charge, e_{eff} is introduced, fitted to the transition $2_1^+ \to 0_1^+$.

The triple reduced matrix element of the quadrupole operator is given in [30]

$$\langle ||| \widetilde{Q}^a ||| \rangle_1 = \sqrt{3}(-1)^\Phi \sqrt{2\widetilde{C}_2(\widetilde{\lambda},\widetilde{\mu})} \text{ with } (-1)^\Phi = \begin{cases} 1 & \text{for } \widetilde{\mu} \neq 0 \\ 0 & \text{for } \widetilde{\mu} = 0 \end{cases} \quad . \tag{4}$$

where the index "1" in the triple reduced matrix element indicates the multiplicity of the coupling $(\widetilde{\lambda}_i, \widetilde{\mu}_i) \otimes (1,1) \to (\widetilde{\lambda}_f, \widetilde{\mu}_f)$.

With this, the reduced $SU(2)$ matrix element is given by

$$\begin{aligned} \langle(\widetilde{\lambda}_f,\widetilde{\mu}_f)\widetilde{\kappa}_f L_f \parallel \widetilde{Q}^a_2 \parallel (\widetilde{\lambda}_i,\widetilde{\mu}_i)\widetilde{\kappa}_i L_i\rangle_1 &= \langle(\widetilde{\lambda}_i,\widetilde{\mu}_i)\widetilde{\kappa}_i L_i, (1,1)1\, 2 \parallel (\widetilde{\lambda}_f,\widetilde{\mu}_f)\widetilde{\kappa}_f L_f\rangle \\ &\quad \langle(\widetilde{\lambda}_f,\widetilde{\mu}_f) ||| \widetilde{\mathbf{Q}} ||| (\widetilde{\lambda}_i,\widetilde{\mu}_i)\rangle_1 \quad , \end{aligned} \tag{5}$$

where L_i and L_f are the phsyical angular momenta.

The $B(E2)$-value is obtained with the formula

$$B(E2; L_i \to L_f) = \frac{2L_f + 1}{2L_i + 1}\langle(\widetilde{\lambda}_f,\widetilde{\mu}_f)\widetilde{\kappa}_f L_f \parallel \widetilde{Q}^a_2 \parallel (\widetilde{\lambda}_i,\widetilde{\mu}_i)\widetilde{\kappa}_i L_i\rangle \quad . \tag{6}$$

In Table 4 the most important $B(E2)$-values are listed. The ground state and the γ-band belong to the same (30,8) irrep, while the β-band is assigned to the (32,4) irrep. This is the reason why there are $B(E2)$ transitions between the ground state and γ-band, while the transition probability within the $\widetilde{SU}(3)$ model is zero and will be suppressed even when $\widetilde{SU}(3)$ breaking interactions are included. The theoretical $B(E2)$-values are compared to experiment in Table 4 and the just explained characteristic of the transition values is clearly seen. In fact, the agreement in general is quite acceptable, considering the simplified model Hamiltonian, but can be improved when the shell mixing terms of the physical quadrupole operator are included and/or $SO(6)$ interaction terms.

4 Conclusions

A semi-microscopic cluster model was applied to the description of ^{162}Dy. This nucleus was chosen as a representative for heavy deformed nuclei in order to illustrate a central apparent problem in nuclear structure physics: In the IBA near the mid shell, the lowest $SU(3)$ irrep has the structure $(2N, 0)$ (N as the number of bosons) and, thus, does not exhibit $B(E2)$-transitions between the γ- and the ground state band, contrary to what is seen in the experiment. It is required to introduce an $\widetilde{SU}(3)$ breaking interaction, for example as done in [8]. We have shown that, taking into account the PEP, this disagreement is resolved because the lowest $\widetilde{SU}(3)$ irrep has two entries $\widetilde{\lambda}$ and $\widetilde{\mu}$ both different from zero. This is a trivial example that long standing problems in structure physics can be resolved straight forwardly, without adding additional degrees of freedam as g-bosons, or similar proposals.

The Semimicroscopic Algebraic Cluster Model for heavy nuclei was able to reproduce the main characteristics of the spectrum and the $B(E2)$ transitions, without recurring to additional interactions. We also emphasized that the IBA does not take into account exchange effects, which leads to the problems in reproducing the transition matrix element. Thus, simply using the correct model space, observing the Pauli Exclusion Principle, can already reproduce the main structure of a nucleus. The present findings may be used to improve the IBA in indicating on how to modify its model space.

A particular result is the description of the ordering of the 0^+ states. The ordering and degeneracies were quite well reproduced in structure and it was shown that this is also due to the presence of the third order Casimir operator. A simple Hamiltonian and knowing he microscopic model space explains already most of the structure of ^{162}Dy.

That the agreement of the $\widetilde{SU}(3)$ model is quite good is probably due to the fact that the ^{162}Dy nucleus is well deformed. However, the present application can be extended directly, first by introducing $\widetilde{SU}(3)$ mixing interactions, as the Casimir operator of $SO(6)$. Note, however, that within the Pauli allowed basis this Casimir operator is not diagonal. Also one can add higher order interactions. We did not do it here, because these extensions, though trivial, only blurres the effects we wanted to highlight.

In future, we will investigate other rare earth nuclei, which show an accumulation of 0^+ states, see for example the talk presented by S. R. Lesher at this symposium. It would be interesting if the $\widetilde{SU(3)}$ is able to describe this intersting phenomenon of accumulation of 0 bands.

Acknowledgment

Financial support from DGAPA-UNAM, project number IN116824 is acknowledged and financial support from the Dept. De la Estructura de la Materia of the ICN-UNAM.

References

[1] P. O. Hess, L. J. Chávez-Nuẽz, Eur. Phys. J. A **57** (2021), 146.
[2] A. Arima, M. Harvey, K. Shimizu, Phys. Lett. B **30** (1969), 517.
[3] K.T. Hecht, A. Adler, Nucl. Phys. A **137** (1969), 129.
[4] D. Bonatsos, I. E. Assimakis, A. Martinou, S. K. Peroulis, S. Sarantopoulou, N. Minkov, Bulg. J. Phys. **30** (2005), 1.
[5] Dennis Bonatsos, Andriana Martinou, S.K. Peroulis, T.J. Mertzimekis, N. Minkov, Symmetry **15** (2023), 169.
[6] A. Leviatan, P. Van Isacker, Phys. Rev.Lett. **89** (2002), 222501.

[7] A. Aprahamian, X. Wu, S.R. Lesher, et al., Nucl. Phys. A **764** (2006), 42.
[8] R. Bijker, J. M. Ruiz, EPJ Web of Conferences **301** (2024), 02001.
[9] A. Aprahamian, C. Casarella, S. R. Leshet, et al., Eur. Phys. J. A **60** (2024), 83.
[10] F. Iachello and A. Arima, *The Interacting Boson Model*, (Cambridge Univeristy Press, Cambridge, England, 1987!.
[11] P. O. Hess, Symmetry **15** (2023) 1197.
[12] O. Castaños, A. Frank, P. O. Hess, M. Mpshinsky, Phys. Rev. C **24** (1981), 1367.
[13] H. Yépez-Martínez, J. Cseh, P. O. Hess, Phys. Rev C **74** (2006), 024319.
[14] P. O. Hess, Eur. Phys. J. A **54** (2018), 32.
[15] P.O. Hess, J.R.M. Berriel-Aguayo1,L.J. Chávez-Nuñez, Eur. Phys. J. A **55** (2019), 71.
[16] J. R. M. Berriel-Aguayo, Peter O. Hess, Symmetry **12** (2020), 738.
[17] P. O. Hess, Symmetry **15** (2023), 1197.
[18] O. Castaños, P. O. Hess, J. P. Draayer, P. Rochford, Nucl. Phys A **524** (1991), 469.
[19] J.P. Draayer, et al., SU(3)-package Routines, Louisiana, USA, 2002.
[20] C. Bahri, D.J. Rowe, J.P. Draayer, Comput. Phys. Commun. **159** (2004), 121.
[21] .Wildermuth,Y.C. Tang, *A Unified Theory of the Nucleus*, (Friedr. Vieweg und Sohn Verlagsgesellschaft mbH, Braunschweig, 1977).
[22] P. Möller, J.R. Nix, W.D. Myers, W.J. Swiatecki, Atom. Data Nucl. Data Tables **59** (1995), 185.
[23] P. Ring, P. Schuck, *The Nuclear Many-Body Problem*, (Springer, Heidelberg, 1980).
[24] O. Castaños, J.P. Draayer, Y. Leschber, Ann. Phys. **180** (1987), 290..
[25] K. T. Hecht, Notas de Física, *Proceedings of the eigth Oaxtepec Simposium on nuclear physics*, (1985), 165.
[26] K. T. Hecht, Nucl. Phys. A **484** (1988), 61.
[27] J. P. Draayer, in *Contemporary Concepts in Physics* Vol. 6, *Algebraic Approaches to Nuclear Structure: Intercating Boson and Fermion Models*, Chapter 7, (harwood academic publishers, USA, 193) p.551.
[28] J. N. Ginocchio, Ann. Phys. **126** (1980), 234.
[29] National Nuclear Data Center, https://www.nndc.bnl.gov/
[30] J. Escher, J.P. Draayer, J. Math. Phys. **39** (1998), 5123.

EPJ Web of Conferences **333**, 01005 (2025) https://doi.org/10.1051/epjconf/202533301005
XLVI Symposium on Nuclear Physics 2025

Elastic Scattering of the ^{6}Li + ^{27}Al system at energies near the Coulomb barrier: From Angular Distributions to Geometric Parameters

X.A. Mendoza-Arriaga[1], J.C. Morales- Rivera[1,*], E. Martinez-Quiroz[2] and E.F. Aguilera[2]

[1]Facultad de Ciencias, Universidad Autónoma del Estado de México, Instituto Literario 100, Código Postal 50000, Toluca, México.

[2]Departamento de Aceleradores y Estudio de Materiales, Instituto Nacional de Investigaciones Nucleares, Apartado Postal 18-1027, Código Postal 11801, México, D.F., México.

Abstract. In this study, we present the experimental elastic scattering angular distributions for the ^{27}Al(^{6}Li,^{6}Li) reaction at energies near the Coulomb barrier: 8.5, 9.0, and 10.0 MeV. The experiment was carried out at the 6 MV Tandem Van de Graaff Accelerator Laboratory of the National Institute for Nuclear Research (ININ) in Mexico. Theoretical calculations were performed using the FRESCO code, and its search version SFRESCO, considering the Woods-Saxon (W-S) form factor and the São Paulo Potential (SPP) to model angular distributions. Additionally, an approach based on the Extended Optical Model was employed to extract optical model parameters, revealing systematic trends in the real and imaginary components of the potential. These trends indicate an increase in absorption effects at higher energies, consistent with previous studies. The results contribute to a better understanding of the Breakup Threshold Anomaly (BTA), a phenomenon associated to weakly bound nuclei. As a future direction, we plan to perform a more detailed analysis of the BTA and its influence on nuclear absorption, and we will extend this approach to other energy regimes.

1 Introduction

Nuclear reactions with heavy ions at energies near the Coulomb barrier were highly studied for their importance to understand the mechanisms of nuclear interactions and the structural proprieties of nuclei. Direct reactions, such as elastic scattering and nucleon transfer, offer significant insights into effective nuclear parameters and geometric properties, including nuclear radius and diffuseness [1]. Many studies have reported the elastic scattering of the ^{6}Li+^{27}Al system [2–4], for the experimental accessibility and the importance to explore weakly bound effects in ions. These previous studies are an important backup and framework for our analysis.

The aim of this work is the analysis of the ^{6}Li+^{27}Al system which exhibits interesting characteristics about the weakly bound nature especially of the ^{6}Li projectile, which can

*e-mail: jmoralesriv@uaemex.mx

probably breakup to $\alpha + d$ [5]. At energies near the Coulomb barrier, these systems exhibit the BTA, where absorption effects caused by the projectile breakup significantly change the reaction dynamics [6]. The BTA can be interpreted as anomalous energy behavior near the Coulomb barrier, that affects the accuracy of the reaction models and provides information about nuclear interactions in low energy heavy ion collisions.

This study is centered to determinate the angular distribution of elastic scattering and present a theoretical analysis identifying the parameters of the nuclear interaction potential which could have a good approximation with the experimental results. With this purpose we analyze experimental data at energies of 8.5, 9.0, and 10.0 MeV, in the laboratory frame, using an approach based on the Extended Optical Model (EOM) as suggested in [7].

To reach this, we used the FRESCO code and its search version SFRESCO [8], which is a very used tool in the computational analysis of nuclear reactions. To describe the interaction between both nuclei in this system we employed theoretical models which describe the absorption effects and some modifications of the optical potential such as the projectile breakup. Phenomenological approaches and the extended optical model provide a great framework for determining effective parameters to describe the nuclear interaction. This method shows information about nuclear radii, surface diffuseness and the coupling effects in reaction dynamics.

2 Experimental setup

To measure elastic scattering, $^{6}Li^{3+}$ ions were accelerated using the Tandem Van de Graaff accelerator facility at ININ to beam energies of 8.5, 9.0, and 10.0 MeV. These ions were directed at an aluminum foil with a thickness of 2.28 μm, with a typical uncertainty of about 4.8% estimated from standard energy loss behavior of alpha particles in Al. The scattered ions were detected by six surface barrier silicon (SSB) detectors labeled A-F mounted on a rotating disk, allowing measurements to be taken in two stages for each beam energy. Detector A was fixed near the chamber wall, maintaining the same angle in both setups, while the remaining detectors were positioned at different angles relative to the incident beam, with slight variations depending on the experimental configuration. In the first configuration, the angles were 25°, 40°, 75°, 130°, 150°, and 170°, respectively. In the second configuration, they were adjusted to 25°, 50°, 85°, 120°, 140°, and 160° in the laboratory frame. A 3×3 mm^2 entrance collimator was used, and each detector was equipped with a 5 mm diameter collimator. The effective depletion thickness of the detectors was 500 μm. Figure 1 illustrates the experimental setup used. During data acquisition, an average beam current of approximately 30 nA was recorded on the target.

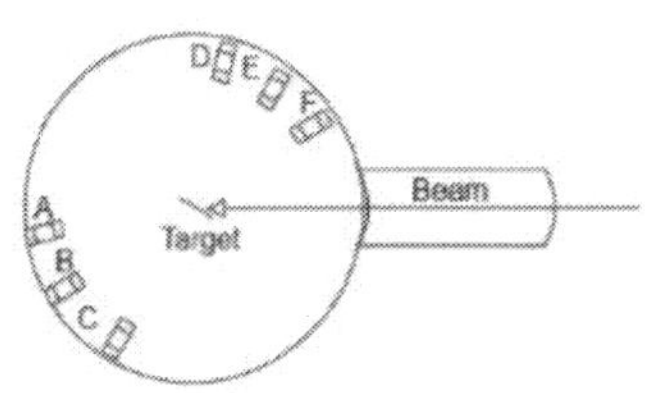

Figure 1: Experimental arrangement used to measure the elastic scattering. A-F represent the SSB detectors.

3 Theoretical analysis

An approximation is made with the Extended Optical Model (EOM) [9] which incorporates additional factors to account for absorption effects. The potential is defined as:

$$U_{EOM} = V_{\text{bare}} + V_C - [iW_{int} + U_F + U_D], \tag{1}$$

Here, V_{bare} is the real nuclear potential (a SPP potential was used), V_C is the Coulomb potential, and W_{int} is the volume imaginary potential that simulates absorption into channels not explicitly treated. The complex polarization potentials U_F and U_D represent the absorption to the fusion and direct reaction channels, respectively, the latter being associated with breakup processes. The terms W_{int} and U_F adopt the standard Woods–Saxon form, while U_D is taken as the derivative of a Woods–Saxon shape, commonly used to describe surface interactions. The radius (r_0) and diffuseness (a) parameters define the geometry of each component and reflect the density distribution of the nucleus, with values adjusted independently for each potential.

The experimental differential elastic scattering cross sections were normalized to the Rutherford elastic scattering to determine the angular distribution. To validate our experimental data a comparison was made between the angular distribution obtained in this work at 10 MeV and the reported by [2, 3]. In Figure 2 we can observe that the data obtained in this study are in a good agreement with previous reports in the literature and in addition measurements are added in a wider range of angles that follow the trend of the other angles. Figure 3 shows the measured elastic scattering angular distributions at the different energies used, these were selected due to their proximity to the barrier and therefore could provide more information on the interaction potentials that are also related with effects like the Breakup Threshold Anomaly (BTA).

To determine the values of the geometric parameters a_D and r_D, a transformation from the scattering angle $\theta_{c.m.}$ to the reduced distance of closest approach d is performed, following the methodology introduced in [7]. In this approach, this representation allows us to relate the angular distributions to the geometry of the interaction and to extract meaningful geometric parameters such as the parameters a_D and r_D. The distance of closest approach D in the Rutherford trajectory is defined as:

$$D = d\left(A_p^{1/3} + A_t^{1/3}\right) = \frac{Z_p Z_t e^2}{2E_{c.m.}}\left(1 + \frac{1}{sin^2\left(\frac{\theta_{c.m.}}{2}\right)}\right) \tag{2}$$

Here, D is the closest approach distance and d is the corresponding reduced distance, Z_p and Z_t are the atomic numbers of the projectile and target, A_p and A_t their mass numbers, and $E_{\text{c.m.}}$ is the energy in the center of mass frame. Angle $\theta_{\text{c.m.}}$ is the scattering angle in the center of mass frame, and d is the reduced distance. Following the methodology introduced in [7] Figure 4 shows the angular distributions as a function of the reduced distance of closest approach $\theta_{c.m.} \to d$. This alternative representation highlights the dynamics of the nuclear interaction. The solid green line is an auxiliary curve without physical meaning that is used to model the drop in the ratio $d\sigma/d\sigma_R$, as a function of the reduced distance d. Defined by three parameters, two adjustable (k, d_c) and one that remains fixed by $b = 1$, see relationship (3), and is used to determine the characteristic distance of closest approach associated with direct reactions. Experimentally, it has been observed that the direct reaction channel dominates over other channels when the ratio $d\sigma/d\sigma_R$ reaches the value of 0.25. This criterion defines the strong absorption distance as $d_S = 1.64$ fm. The quantity d_s represents the reduced distance of maximum approach characteristic of direct reactions, i.e., $g(d_s) = 0.25$. In this context,

d_s is used as the value for r_D. The dashed curve corresponds to a two parameter function of the form $1 - \gamma f'(r)$ that fits over $g(d)$, where γ it is a constant determined numerically to achieve the best possible fit to the data or to the target function and $f'(r)$ is the derivative of a volume W–S form factor. This surface form is used to determine the diffuseness parameter, with $a_D = 0.98$ fm being the value obtained from the fit.

$$g(d) = \frac{b}{1 + e^{-k(d-d_c)}} \quad (3)$$

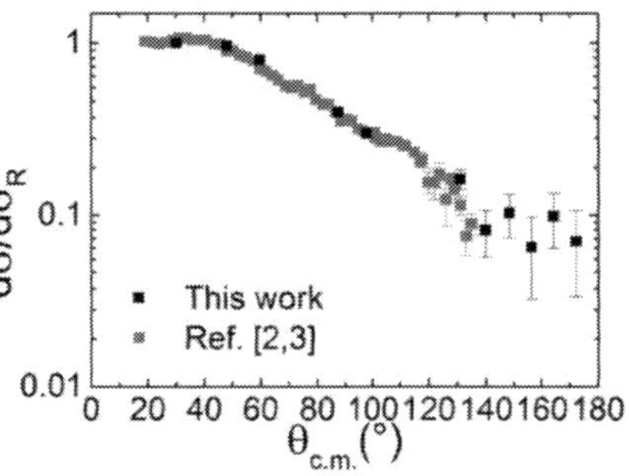

Figure 2: **(Color online)** Comparison of the measured data in this work with previously published data at 10 MeV by [2, 3].

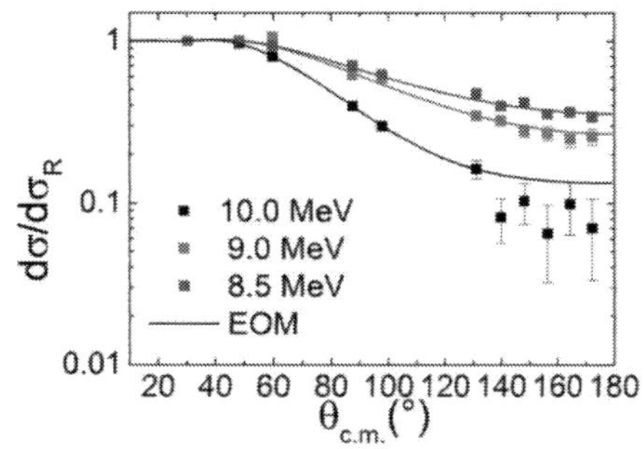

Figure 3: **(Color online)** Experimental elastic scattering angular distributions for the $^{6}Li+^{27}Al$ system. Blue, red and black symbols represents the beam energies of 8.5, 9.0 and 10 MeV, respectively. Solid lines correspond to FRESCO calculations.

Here, $g(d)$, is used to approximate the transition region in the absorption profile as a function of d to extract geometric parameters. The fitting of the experimental elastic scattering data was performed by optimizing the strengths V_D,V_F,W_D, and W_F of the polarization potentials with SFRESCO. The parameters used in the calculations are those suggested in the literature [7, 9] are as follows: $W_{int} = 50$ MeV $r_0 = 1.0$ fm, $a_0 = 0.20$ fm were kept fixed for all energies. Furthermore, real and imaginary part of the volume and surface polarization potentials were also fixed as $r_F = 1.40$ fm, $a_F = 0.43$ fm, $r_D = 1.64$ fm and $a_D = 0.98$ fm. The comparison between the experimental data (symbols) and the calculations (lines) made by means of the EOM is shown in Figure 3. We can see that our fits are in agreement with the experimental data, showing a slight discrepancy at large angles at higher energy.

In table 1 we report the optimized parameters V_D,V_F,W_D, and W_F obtained by SFRESCO. Additionally, using EOM the fusion cross section is obtained from the absorption with W_{int} which represents the absorption inside the barrier, and W_F, which incorporates the absorption below the barrier [9]. To determine the fusion cross sections at the studied energies, an interpolation is performed with previously reported experimental data [10] using Wong's equation with three adjustable parameters, see equation (4).

$$\sigma(E_{c.m.}) = \frac{\hbar\omega R_B^2}{2E_{c.m.}} Ln\left(1 + e^{\frac{2(E_{c.m.}-B)}{\hbar\omega}}\right), \quad (4)$$

Table 1: Optimized parameters V_D,V_F,W_D, and W_F and fusion cross section obtained by SFRESCO. Geometric parameters were kept constant; see text for details.

Effective energy	V_D(MeV)	V_F(MeV)	W_D(MeV)	W_F(MeV)	$\frac{\chi^2}{N}$	σ_{fus}(mb)
6.95	0.21	13.45	0.41	1.35	10.90	117.73
7.36	0.05	8.50	0.30	4.58	7.58	157.38
8.18	-0.11	6.91	0.38	1.55	5.00	465.21

where $E_{c.m.}$, R_B, B, and $\hbar\omega$ define the center of mass energy, radii, height, and width of the Coulomb barrier respectively. The fusion cross sections obtained after optimization of the parameters are compared with reported experimental values and the Wong's equation.

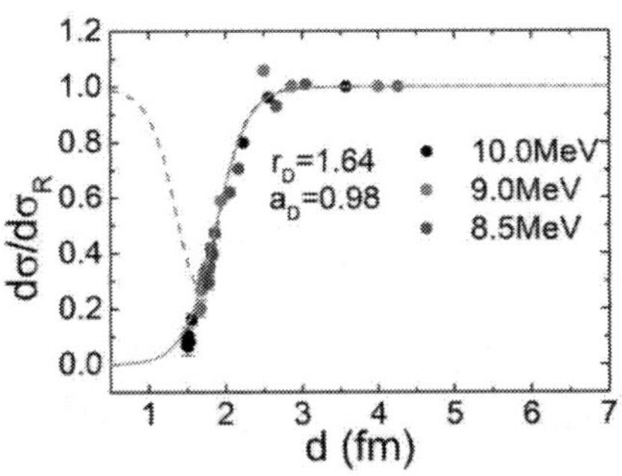

Figure 4: **(Color online)** Plot of $d\sigma/(d\sigma_R)$ vs d for the elastic scattering data. The meaning of the curves is detailed in the text.

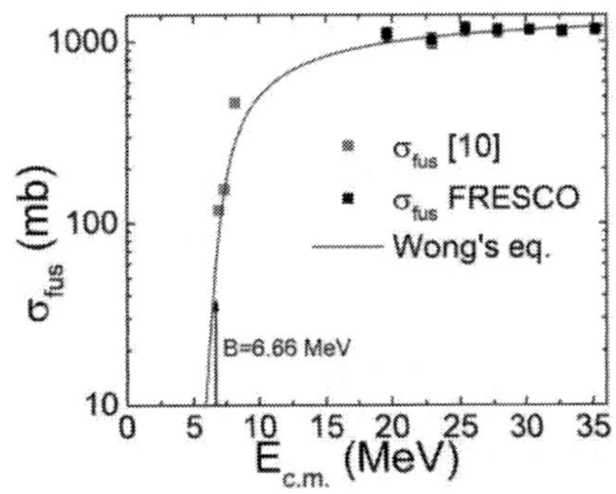

Figure 5: **(Color online)** Comparison of the fusion cross sections calculated using SFRESCO with experimental data obtained from [10]. The curve represent the fusion excitation function using the Wong's equation.

The theoretical analysis presented provides a framework to understanding the dynamics in elastic scattering and fusion of the $^{6}Li+^{27}Al$ system with energies near Coulomb barrier. Using the EOM and the Wong's equations we obtained the geometric parameters necessary to characterize the interaction. Comparisons between theoretical analysis and experimental data validated the EOM approach and highlight the importance of surface interactions and reaction channels on nuclear dynamics.

4 Conclusions

In general, the results of the theoretical analysis with the EOM have similarities with the experimental data but with some differences at larger angles which ones possible are related with the BTA and its influence in the reaction channels, particularly in direct reactions increasing the absorption. Besides, the parameters determined for the EOM showed expected trends such as an increase in the imaginary component of the potential with energy. This behavior agrees with what was reported in previous studies on weakly bound projectiles and reaction mechanisms. If we compared the EOM approach with previous analysis of ^{6}Li + ^{27}Al systems using the standard optical models [2–4], we can observe that the EOM provides a more explicit way of observing the contributions of fusion and breakup channels, therefore their separation allows a more focused analysis of these channels that are the most influential in the BTA.

The transformation of the angular distribution to the reduced distance of closest approach provides a way to determine and refine the system's geometric characteristics. Furthermore, the results of this work contribute to the development of basic science that helps in the understanding of nuclear reactions with weakly bound nuclei and provides a basis for future studies, such as a detailed analysis of the influence of BTA, and it is planned to use these results as a basis and add an analogous analysis at more energies around the barrier.

References

[1] A. Diaz-Torres, et al.,How does breakup influence the total fusion of 6, 7Li at the Coulomb barrier?. Phys. Rev. C **68**, 044607 (2003). https://dx.doi.org/10.1103/PhysRevC.68.044607

[2] J. M. Figueira, et al., Breakup threshold anomaly in the elastic scattering of 6Li on 27Al. Phys. Rev. **75(1)**, 017602 (2007). https://doi.org/10.1103/PhysRevC.75.017602

[3] J. O. Fernández Niello, et al., Study of the Threshold Anomaly in the Scattering of Li Isotopes on 27Al. Nucl. Phys. A **787**, 484c-490c. (2007). https://doi.org/10.1016/j.nuclphysa.2006.12.072

[4] S.H. Hamada et al., Extensive investigation of the 6Li + 27Al and 6He + 27Al systems using different models. Phys. Scr. **97**, 125303 (2022). https://doi.org/10.1088/1402-4896/ac9dc9

[5] M. S. Hussein, et al., Low Energy Transfer Cross-Section for Borromean Halo Nuclei and the Breakup Threshold Anomaly. AIP Conf. Proc. C **853**, 378–383 (2006). https://doi.org/10.1063/1.2338405

[6] M. Mazzocco, et al., Elastic scattering for the 8B and 7Be + 208Pb systems at near-Coulomb barrier energies. Phys. Rev. C **100**, 024602 (2019). https://doi.org/10.1103/PhysRevC.100.024602

[7] F. Torabi, et al., Systematic study of elastic scattering and fusion induced by weakly bound 6Li on medium mass targets. Threshold anomalies. Nucl. Phys. A **994**, 121661 (2020). https://doi.org/10.1016/j.nuclphysa.2019.121661

[8] I. J. Thompson, Coupled reaction channels calculations in nuclear physics. Comput. Phys. Rep. **7**, 167 (1988). https://doi.org/10.1016/0167-7977(88)90005-6

[9] E. F. Aguilera, et al., Simultaneous analysis of elastic scattering and fusion in 6He+64Zn: A transition in direct reaction mechanisms, striking threshold anomalies, and halo effects. Phys. Rev. C **104**, 054612 (2021). https://doi.org/10.1103/PhysRevC.104.054612

[10] I. Padron, et al., Fusion of stable weakly bound nuclei with 27Al and 64Zn. Phys. Rev. C **66**, 044608 (2002). https://doi.org/10.1103/PhysRevC.66.044608

EPJ Web of Conferences **333**, 01006 (2025)
XLVI Symposium on Nuclear Physics 2025
https://doi.org/10.1051/epjconf/202533301006

Experimental Determination of the Neutron Scatter Contribution to the Neutron Reference Field of the SSDL-ININ

B. Navarro-Hurtado[1,2], *M.* Maldonado-Velázquez[2,*], *H.* Mendoza-Nava[2], and *J.C.* Morales-Rivera[1]

[1]Facultad de Ciencias, Universidad Autónoma del Estado de México, Instituto Literario 100, Toluca, C.P. 50000, México.
[2]Centro de Metrología de Radiaciones Ionizantes, Instituto Nacional de Investigaciones Nucleares, Carretera México-Toluca S/N, La Marquesa, Ocoyoacac, C.P. 52750, México.

Abstract. Room neutron scattering was measured at the SSDL-ININ neutron calibration room, by using a secondary standard ^{241}AmBe source and a Berthold D7547 Wildbad detector placed in a 25.4 mm moderator. The contribution of neutron scattered radiation was determined through two methods: the shadow-cone method, consisting in the placement of a shadow-cone in the direct path between the source and the detector; and the semiempirical method, a mathematical formulation, based on the assumption that the scattered contribution of the neutron spectrum can be estimated from the inverse-square law. Additionally, Monte Carlo calculations were made to simulate the transport of neutrons in the room, obtaining results consistent with those experimentally determined.

1 Introduction

Neutron detection is a challenging subject due to their non-Coulomb interactions with matter. One of the most critical issues in neutron metrology is determining the contribution of scattered neutrons to reference fields, which increases the neutron fluence rate inside the measurement room. These neutrons are primarily scattered from the floor, walls, ceiling, and other structures. Therefore, it is essential to assess the scattered neutron contribution for neutron metrology rooms; according to the ISO 8529-2 [1], the scattered neutron contribution to the reference field for a neutron room should not exceed 40%. This paper presents the results of the measured and simulated scatter contribution to the neutron reference field of the Secondary Standard Dosimetry Laboratory (SSDL) at the National Institute for Nuclear Research (ININ), taking into account the specific conditions of the facility in Mexico.

2 Materials and methods

2.1 Neutron calibration room and instruments

The neutron calibration room at the SSDL-ININ has an octagonal geometry with an apothem of 4 m and a height of 5.5 m. It has a false floor made of metal grating, beneath which there is a 3 m deep basement. Four walls are made out of concrete for shielding, and the remaining four are made of thin iron sheets to minimize neutron scattering (see Figure 1). Due to the high altitude (~3000 m above sea level), environmental conditions were monitored during the measurements, finding an average temperature of 18.9 °C, an atmospheric pressure of 710.47 hPa and an average relative humidity of 58.5%, corresponding to an air density of 8.42×10^{-4} g/cm^3.

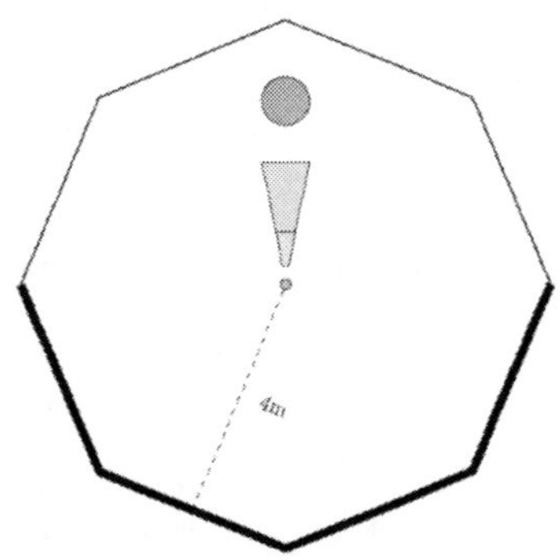

Figure 1. Diagram of the octogonal neutron room (top view) with four concrete walls (thick lines), four thin iron sheets (thin lines) and alignment of instruments.

A calibrated ^{241}AmBe neutron source was placed at the center of the room. The neutron detector used was a ^{6}Li(Eu) crystal with a diameter of 10 mm, and height of 2 mm, within a 25.4 cm diameter spherical moderator. The detector was connected to modular electronic instruments in order to optimize the data acquisition. All measurements were conducted over counting times sufficient to accumulate at least 10,000 counts for each experimental point.

[*]e-mail: marissa.m.servicios@inin.gob.mx

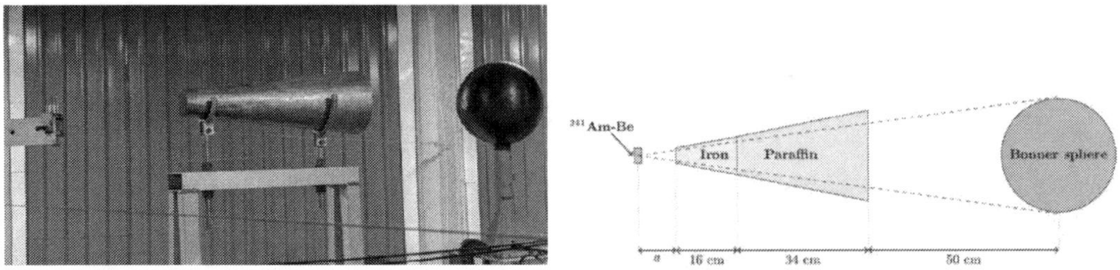

(a) Experimental setup used for the shadow-cone method, with the source, shadow-cone and spherical detector.

(b) Diagram of the experimental setup for the shadow-cone method measurements. Distance a took values of 10, 20 and 30 cm, for each of the three distances from center of the source to center of the detector.

Figure 2. Side by side comparison between the experimental arrangement and the experimental design of the shadow-cone method.

2.2 Shadow-cone method

According to [2], the shadow-cone method is useful to experimentally determine neutron scattering contributions to a reference field. The method consists in the placement of a shadow-cone in the direct path between the source and the detector in order to shadow the direct neutrons coming from the source, thereby allowing only the scattered neutrons to be measured; measurements are also made without shadow-cone to determine the total amount of neutrons at each measurement point, Figure 2(a) shows the experimental setup used for this method.

The contribution of direct neutrons to the detector reading is obtained by subtracting the measurement with the shadow cone $M_S(l)$, from the measurement without the shadow cone, $M_T(l)$. This result must then be corrected for air outscatter $F_A(l)$, which accounts for neutrons that come from the source directly to the detector, but are scattered by air before reaching the moderator. Moreover, in this study air attenuation correction considers the specific air conditions at SSDL, by equations (1) and (2),

$$F_A(l, E) = \exp[l\Sigma(E)] \quad (1)$$

$$F'_A(l, E) = F_A(l, E)^{\rho_{SSDL}/\rho_{std}} \quad (2)$$

where $F'_A(l, E)$ represents the air attenuation correction for non-standard air conditions, ρ_{SSDL} is the air density at SSDL, and ρ_{std} is the air density under standard conditions. The air densities were calculated using the revised formula for density of moist air, as given in [3]. A linear attenuation coefficient $\Sigma(E) = 833 \times 10^{-7}\text{cm}^{-1}$ was used for the ^{241}AmBe reference source.

The relationship between the measured direct neutrons and the neutrons emerging from the source is described by equation (3).

$$[M_T(l) - M_S(l)]\, F'_A(l) = \frac{k}{l^2} \quad (3)$$

with

$$k = R_\Phi B_\Omega \quad (4)$$

$$B_\Omega = \frac{BF_1(\theta)}{4\pi} \quad (5)$$

Where k is the characteristic constant of the source-detector arrangement, R_Φ is the fluence response and B_Ω is the certified angular strength of the source with B as the total strength. The term $F_I(\theta)$ in equation (5) corrects for source anisotropy, as the strength of the neutron source varies with direction of emission. The reference source used in the experiment has an anisotropy correction factor of 1.0248 for the direction $\theta = 90^o$.

A total of six measurements were conducted, three with the shadow-cone in place and three without it, at distances of 110 cm, 120 cm and 130 cm from the center of the source to the center of the detector. These distances l were chosen to be greater than twice the lenght of the shadow-cone, ensuring that the geometric correction factor $F_1(l)$ is practically equal to one [2]. The shadow cone used in this study was 50 cm long, and it was composed of two materials, 16 cm of iron and 34 cm of paraffin wax, with a front end diameter of 6.66 cm and a rear diameter of 17.26 cm as shown in Figure 2(b).

2.3 Semi-empirical method

The semi-empirical method determines the contribution of scattered neutrons through a mathematical formulation based on the assumption that the scattered component can be derived from the inverse-square law. This involves taking multiple measurements of the total neutron field $M_T(l)$ at different distances l. These measurements are then corrected and fitted to determine the fluence response R_Φ of the measuring device and the fractional room-scatter contribution S to the neutron field. The model is described by equation (6).

$$\frac{M_T(l)}{\varphi F_1(l)(1 + Al)} = R_\Phi(1 + Sl^2) \quad (6)$$

with

$$F_1(l) = 1 + \delta\left(\frac{r_D}{2l}\right)^2 \quad (7)$$

$$\varphi = \frac{B_\Omega}{l^2} \tag{8}$$

Equation (7) is referred to as the geometric correction, and takes into account the increase in the detector's response by a factor of $(1 + \delta G)$ above that expected according to the inverse-square law. G is the fraction of additional neutrons entering the volume of the polyethylene sphere, and the effectiveness parameter δ takes into account the probability of these extra neutrons in producing a response in the detector [4]. Equation (8) is the free field fluence rate to which the detector has been exposed at a distance l.

For this method, 22 measurements were taken at different distances ranging from 26 cm to 130 cm. The left side of equation (6) was applied to every measure and then plotted against l^2. This plot was fitted to a linear function, through the weighted least squares method, to obtain the fluence response R_Φ of the detector and the fractional room-scattered component of the field S [5].

For the data analysis the standard uncertainties considered were those corresponding to the distance l and the number of counts $M_T(l)$, as well as the corresponding to the air attenuation factor, for which the linear attenuation coefficient $\Sigma(E)$ and the moist air density ρ_{SSDL} were taken into account. From the certificates of the source, the source strength and anisotropy uncertainties were also considered. For geometry corrections, the uncertainties of the effectiveness parameter and the radius of the detector were combined.

The standard uncertainties of the results were determined based on [6].

3 Monte Carlo simulations

Monte Carlo simulations were carried out using OpenMC [7] to reproduce the experimental data. The geometry of the neutron room was modeled based on its actual dimensions, and the Compendium of Material Composition Data for Radiation Transport Modeling [8] was used to model the materials for the walls, floor, ceiling and instruments. It is also important to mention that the thermal neutron scattering data $S(\alpha,\beta)$ for the moderator materials was included in the simulation.

For the instrumentation section, we defined a cylindrical neutron source with a diameter of 2.12 cm and a length of 2.5 cm, with a strength corrected by radioactive decay of 9.0036×10^6 s^{-1} and an energy spectrum obtained from [9]. The detector was modeled as a 25.4 cm spherical cell made of polyethylene with a scintillator positioned in the center of the sphere, with a density of 3.494 g/cm^3. The scintillator's composition was 4.36% of ^{6}Li, 0.18% of ^{7}Li, and 95.46% of ^{127}I [10]. Finally, for the shadow-cone two conic cells were defined, one filled with iron and one with paraffin wax.

Experimental setups with and without shadow-cone were reproduced, hence 6 simulations were ran with 2×10^9 histories each, to reach acceptable standard deviations.

4 Results and discussion

The experimental results for the semi-empirical method at the 22 different positions and the weighted linear fit applied are shown in Figure 3. The parameters of the linear fit were used to obtain, from equation (6) the, fractional room scatter component S and the instrument's fluence response R_Φ. Results are shown in Table 1.

Table 1. Obtained coefficients from the weigthed least-squared fit, where u(x) represents the uncertainty of each parameter.

Parameters			
R_ϕ	u(R_ϕ)	S	u(S)
0.17	5.38×10^{-4}	1.13×10^{-5}	5.45×10^{-7}

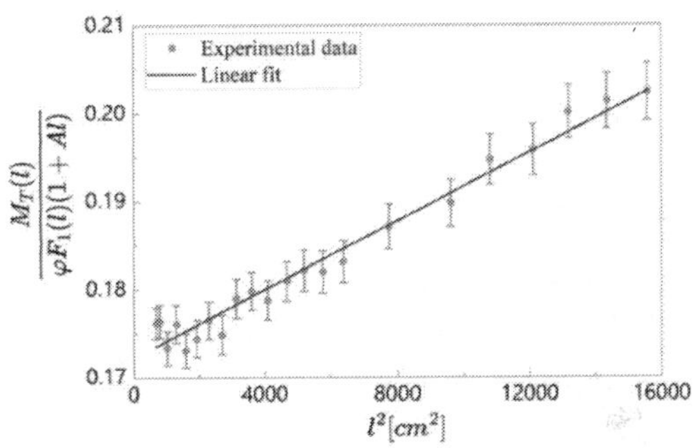

Figure 3. Obtained results for the 22 distances l of the left side of equation (6) with uncertainties, and the weighted least-squared fit applied to these data.

To compare the shadow-cone method results with the corresponding semi-empirical method, the fractional room scatter contribution S obtained from the linear fit was used to calculate the scattered contribution at each one of the three distances used in the shadow-cone method. The comparison showed a good agreement between both methods, with differences within the combined uncertainties; results are shown in Table 2; notice that the quantity S is dimensionless.

The plotted results of this study are shown in Figure 4. Simulated points are consistently lower than the corresponding measured points. These differences can be explained by the presence, in the room, of instruments and structures yet to be included in the geometry model. It should be noted that the uncertainty of the MC simulation is merely statistical, and does not include other contributions, such as those due to uncertainty of nuclear data and material composition.

Table 2. Comparison between experimental and simulated results with uncertainties $u(S)$.

	Shadow-cone method		Semi-empirical method		Monte Carlo Simulations	
Distance [cm]	Scattering (S)	u(S)	Scattering (S)	u(S)	Scattering (S)	u(S)
110	0.1346	0.0026	0.1364	0.0066	0.1062	0.0037
120	0.1597	0.0029	0.1623	0.0079	0.1319	0.0048
130	0.1925	0.0025	0.1905	0.0092	0.1467	0.0053

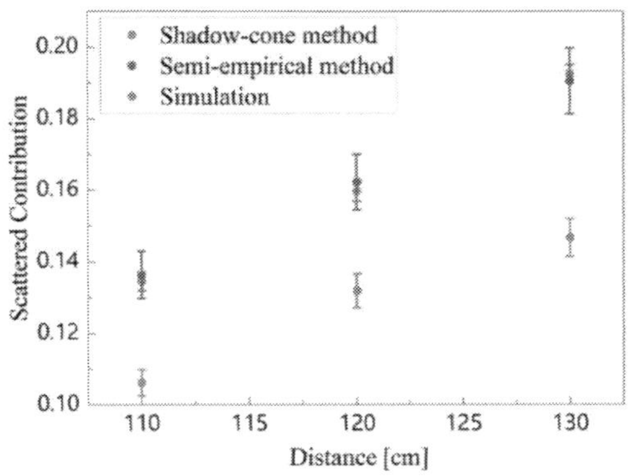

Figure 4. Comparison of the obtained results by means of the Shadow-cone and Semi-empirical methods with those obtained from the Monte Carlo Simulations for the three distances considered.

5 Conclusions

Scattered neutron contribution to the reference field of the SSDL-ININ was determined through two experimental methods, obtaining consistent results, within uncertainties. A fractional room-scattered component S of 1.13×10^{-5}, with a standard uncertainty of 5.45×10^{-7} was found for the instrument. At the larger distance, a scatter contribution to the reference field below 20% was found, which complies with the requirements given in [1]. Simulations results showed tendencies in good agreement with measurements, however, they are consistenly lower for all experimental points. This difference can be explained with the presence of some other items in the room during the experiment. Therefore, it is concluded the consistency of both methods for determining the neutron scattering contribution together with the Monte Carlo simulations in this study.

References

[1] I.O. for Standardization, ISO 8529-2. Reference neutron radiations — Part 2: Calibration fundamentals of radiation protection devices related to the basic quantities characterizing the radiation field (International Organization for Standardization, 2000)

[2] N. Mirzajani, R. Ciolini, G. Curzio, Analysis of the application of the shadow cone technique for the determination of neutron spectrum with bonner sphere spectrometer, Nuclear Instruments and Methods in Physics Research Section A: Accelerators, Spectrometers, Detectors and Associated Equipment **722**, 24 (2013). https://doi.org/10.1016/j.nima.2013.04.030

[3] A. Picard, R.S. Davis, M. Gläser, K. Fujii, Revised formula for the density of moist air (cipm-2007), Metrologia **45**, 149 (2008). 10.1088/0026-1394/45/2/004

[4] J.B. Hunt, The calibration of neutron sensitive spherical devices, Radiation protection dosimetry **8**, 239 (1984). 10.1093/oxfordjournals.rpd.a083060

[5] S.K. Lee, S.I. Kim, J. Lee, I. Chang, J.L. Kim, H. Kim, M.C. Kim, B.H. Kim, Evaluation of neutron scattering correction using the semi-empirical method and the shadow-cone method for the neutron field of the korea atomic energy research institute, Radiation Protection Dosimetry **180**, 46 (2018). 10.1093/rpd/ncx219

[6] BIPM, IEC, IFCC, ILAC, ISO, IUPAC, IUPAP, OIML, Evaluation of measurement data — Guide to the expression of uncertainty in measurement (2008). https://doi.org/10.59161/JCGM100-2008E

[7] The openmc monte carlo code — openmc documentation, version 0.15.0, `https://openmc.org/`

[8] R.G. Williams, Compendium of material composition data for radiation transport modeling, PNNL-15870, Pacific North National Laboratory (2011). 10.2172/1023125

[9] H. Kluge, K. Weise, The neutron energy spectrum of a 241am-be(alpha,n) source and resulting mean fluence to dose equivalent conversion factors, Radiation Protection Dosimetry **2**, 85 (1982). 10.1093/oxfordjournals.rpd.a080571

[10] H.R. Vega-Carrillo, I. Donaire, E. Gallego, E. Manzanares-Acuña, A. Lorente, M.P. Iñiguez, A. Martin-Martin, J.L. Gutierrez-Villanueva, Calculation of response matrix of a bss with 6lii scintillator, Revista mexicana de física **54**, 57 (2008).

EPJ Web of Conferences **333**, 01007 (2025) https://doi.org/10.1051/epjconf/202533301007
XLVI Symposium on Nuclear Physics 2025

Role of Absorption Inside the Barrier For Different Weakly Bound Systems

S. Gomez-Rivera[1,2,*], *E. F.* Aguilera[1,**], *J. C.* Morales-Rivera[2,***], *E.* Martinez-Quiroz[1], and *D. A.* Godoy[1]

[1]Departamento de Aceleradores y Estudio de Materiales, Instituto Nacional de Investigaciones Nucleares, Apartado Postal 18-1027, Código Postal 11801, México, Distrito Federal, México.
[2]Facultad de Ciencias, Universidad Autónoma del Estado de México, Instituto Literario 100, Código Postal 50000, Toluca, México.

Abstract. This investigation includes an analysis of the effects of the short-range absorptive potential W_{int}, included in the Extended Optical Model Framework (EOM) for several weakly bound systems, namely ^{6}He+^{64}Zn, ^{9}Be+^{28}Si, ^{14}N+^{9}Be and ^{9}Be+^{12}C. In addition to W_{int}, the EOM includes polarization potentials U_D, U_F, associated with direct and fusion couplings, respectively. The geometric parameters were obtained (for U_D), and the optimization of the corresponding strength parameters was carried out for both polarization potentials using the Sfresco code. A parameterization is obtained, which simultaneously reproduces the experimental data of elastic scattering and fusion cross sections, while the dispersion relation is also satisfied. The influence of this absorptive potential W_{int} has been described by analyzing its effect on angular distributions and reflection coefficients, as well as inspecting the semi-classical deflection function.

1 Introduction

In the present work, data for several weakly bound systems will be analyzed with the aim of shedding light on the role that absorption inside the barrier plays on the corresponding elastic scattering process. In order to consistently describe the respective fusion data in addition, the framework of the Extended Optical Model (EOM) will be used. The EOM originally developed by Udagawa, Tamura and Kim [1, 2] uses the following form of the Optical Potential:

$$V(r) = V_C + V_{bare} - [U_D + U_F + i\,W_{int}], \qquad (1)$$

where V_{bare} and V_C are purely real components, $i\,W_{int}$ is purely imaginary, while U_F and U_D are complex. With respect to the standard Optical Model, the EOM's purpose is to facilitate the extraction of polarization potentials U_D (using the surface Woods-Saxon form factor) and U_F (using the volumetric Woods-Saxon form factor) associated with direct and fusion couplings, respectively.

The nuclear potentials that appear in the EOM can be categorized as follows: V_{bare} plays the role of a bare nuclear potential, which is modified by the effect of direct and fusion couplings. The effect of such couplings is modeled by phenomenological polarization potentials U_D and U_F, using superficial and volumetrical form factors of Woods-Saxon type, respectively. In order to satisfy a dispersion relation, both U_D and U_F must be complex, so $U_D = V_D + iW_D$ and $U_F = V_F + iW_F$.

In general, a strongly absorptive short-range potential W_{int} is used to prevent unphysical reflection of lower partial waves which could survive at very small values of r if W_{int} is not introduced. This absorptive potential is parametrized using the volumetric Woods-Saxon form factor [3].

Due to the generally large strength of the bare potential V_{bare} and also because W_F and W_D turn out to be rather weak, reflections of lower partial waves may appear in the asymptotic region, which may cause unphysical oscillations of differential elastic scattering cross sections at large angles, particularly at relatively high energies above the Coulomb barrier. However, physically, such reflections are not normally expected because of the strong absorption that exists inside the nucleus. W_{int} is therefore introduced to simulate such strong absorption, thus eliminating this seemingly unphysical effect.

The need for introducing W_{int} was first discussed in [3–5] and its use has been applied in the analysis of a large variety of systems, namely ^{6}Li + ^{208}Pb [3], ^{7}Li + ^{208}Pb [4], ^{12}C + ^{208}Pb [5], ^{9}Be + (^{28}Si, ^{144}Sm, ^{208}Pb) [6], ^{6}Li + (^{58}Ni, ^{59}Co, ^{64}Ni, ^{64}Zn) [7], ^{6}He + ^{64}Zn [8], ^{9}Be + ^{12}C [9].

However, an interesting feature regarding the behavior of certain systems having a ^{9}Be projectile was recently noticed. While a fairly strong short range absorbing potential W_{int} was apparently necessary in the analysis of the ^{9}Be + (^{28}Si, ^{144}Sm, ^{208}Pb) systems of Reference [6], for the case of the lighter ^{9}Be + ^{12}C system of Reference [9] a null strength for this potential, W_{int} = 0 MeV, was found instead. This was necessary in order to describe specific large angle oscillations seen in the experimental angular distributions corresponding to particular energies; it would seem, at least for ^{9}Be projectiles that light targets such as ^{12}C produce a rather weak internal absorption which actually allows for some partial wave reflections to appear as

*e-mail: sgomezr002@uaemex.mx
**e-mail: eli.aguilera@inin.gob.mx
***e-mail: jmoralesriv@uaemex.mx

actual oscillations in the elastic scattering angular distributions experimentally measured.

To further understand these observations, in the present work a study is performed applying the EOM to four ^{9}Be and ^{6}He induced systems, with target masses ranging from 12 up to 64. To be specific, the ^{9}Be + (^{12}C, ^{28}Si) and ^{6}He + ^{64}Zn systems will be reanalyzed in order to highlight the effect of W_{int} on the respective elastic scattering angular distributions. These systems were previously analyzed in References [6, 8, 9], respectively, but some new elements of analysis will be introduced here. In addition, the rather light ^{14}N + ^{9}Be system will be analyzed here for the first time within the EOM.

As a result, the use of a non-zero W_{int} in the two heavier systems becomes necessary while the corresponding required strength becomes lower for the two lighter systems; in fact, it will be verified that it is indeed necessary to have a null W_{int} in order to better reproduce the data for the lightest analyzed system, ^{9}Be + ^{12}C, and the oscillations observed in the data will be correlated with the corresponding partial waves being reflected.

During this study, it was sometimes necessary to estimate the fusion cross section when the available experimental data did not correspond to the exact bombarding energy for elastic scattering. This is why we have utilized the Wong's equation (2) [10], mainly for interpolation purposes.

$$\sigma(E) = \frac{\hbar\omega R_B^2}{2E_{C.M.}} Ln\left(1 + \exp\left(\frac{2\pi}{\hbar\omega}(E - V_B)\right)\right) \quad (2)$$

On the other hand, equation (1) contains only four free strength parameters (geometrical parameters were obtained following the methodology reported in Reference [8]); two parameters come from the direct polarization potential (V_{0D}, W_{0D}) and two more from the fusion polarization potential (V_{0F}, W_{0F}). These parameters are presented for the four systems after an optimization process, focusing on satisfying the dispersion relation for each system. The dispersion relation corresponds to the next equation [11]:

$$V(E) = V_0(E_s) + \frac{P}{\pi}(E - E_s)\int_0^\infty \frac{W(r, E')}{(E' - E_s)(E' - E)}dE', \quad (3)$$

where P means the principal value of the integral, $V(E_S)$ is the value of $V(E)$ at some reference energy E_S. Considering the particular form of the U_D, U_F potentials, equation (3) reduces to a relation between only the respective strength parameters.

When we compare EOM with other models, such as conventional Optical Model analysis (OM) or Continuum-Discretized Coupled Channels (CDCC), EOM has some advantages. It allows us to study the behavior of the direct and fusion potentials separately. This is useful because we can study both together when they are coupled with the elastic channel. This means that we can assess the Break-up Threshold Anomalies (BTA) and Threshold Anomalies (TA) phenomena at the dispersion relations.

At the same time, EOM eliminates the renormalization process required when weakly bound nuclei are involved, which is related to the dominance of break-up processes in nuclear reactions [12]. Conversely, previous CDCC analyses of weakly bound systems have shown that the two polarization potentials resulting from EOM and CDCC provide the same physical information; namely, their magnitudes and radial dependence are quite similar. However, CDCC studies are unable to reproduce the TA information that EOM can when break-up processes are stronger than other more complicated interactions between nuclei [3, 13].

2 The ^{6}He + ^{64}Zn and ^{9}Be + ^{28}Si systems

In this study, the São Paulo Potential (SPP, [14]) is used for V_{bare}, while Woods-Saxon and derivative Woods-Saxon form factors are used for the fusion and direct potentials, respectively. A Woods-Saxon form factor is also used for the internal potential W_{int} but to properly describe absorption inside the barrier, small values of radius and diffuseness (r_0, a_0 = 1.0, 0.2 fm and r_0, a_0 = 0.8, 0.3 fm, respectively) are used in this case. The codes FRESCO and SFRESCO [15] are utilized to find optimized parameter values that fit the available data for elastic scattering angular distributions and for fusion, simultaneously watching the validity of the dispersion relation for each system. The optimization procedure has been described in detail elsewhere [7].

Table 1. Optimized parameter values for ^{6}He + ^{64}Zn. The first five columns are in MeV. The geometric parameters for both polarization potential were (r_{0D}, a_{0D}) = (1.53, 1.061) fm and (r_{0F}, a_{0F}) = (1.40, 0.43) fm.

E_{lab}	V_{0D}	W_{0D}	V_{0F}	W_{0F}	χ^2	σ_F (mb)
9.84	-0.28	0.41	3.00	4.00	0.336	132.09
13.45	-0.39	0.73	1.96	4.60	0.435	410.55
14.85	-0.53	0.26	1.25	4.30	56.748	748.53
17.93	-0.51	0.29	0.8	4.30	211.348	976.89

The results for the ^{6}He + ^{64}Zn system are presented in Figure 1, where the experimental angular distributions for the four available energies are shown along with the respective EOM calculations corresponding to different values of the strength W_{int}. The values of the respective optimized parameters are given in Table 1. Since the Coulomb barrier for this system is B = 8.42 MeV (equivalent to 9.2 MeV in the laboratory frame), all measured energies are above the barrier in this case.

It can be seen in Figure 1 that using a null W_{int} would not be consistent with the data. For the higher energies, large oscillations would be predicted at large angles if W_{int} = 0 MeV is used in the calculations. However, any strength of 10 MeV or higher would be enough to make the calculations agree with the data for all measured energies. The value W_{int} = 50 MeV used for this system in Reference [8] is thus justified. The χ^2 values reported in Table 1 corresponds to this value of W_{int}, black continuous curves in Figure 1.

Table 2. Optimized parameter values for ^{9}Be + ^{28}Si. The first five columns are in MeV. The geometric parameters for both polarization potential were (r_{0D}, a_{0D}) = (1.50, 0.60) fm and (r_{0F}, a_{0F}) = (1.40, 0.50) fm.

E_{lab}	V_{0D}	W_{0D}	V_{0F}	W_{0F}	χ^2	σ_F (mb)
12	-0.04	0.76	1.10	6.20	0.199	125.21
13	0.11	0.88	0.10	6.90	0.656	208.57
14	0.10	0.95	-1.57	8.20	0.330	286.43
17	0.25	0.95	-0.53	8.32	0.155	525.00
20	0.39	0.98	-3.64	7.33	1.119	641.44
23	0.51	0.90	-3.76	8.64	0.264	821.51
26	0.42	0.88	-4.07	7.18	0.998	881.20
30	0.46	1.04	-4.71	6.98	2.235	924.13

An important difference of the present analysis for ^{6}He + ^{64}Zn compared to that of Reference [8] lies in the choice of geometric parameters (r_D, a_D) for the direct reaction part of the potential. The significant spread out of points in the plot of Figure 1 of Reference [8] leads to an ambiguity in the values of these parameters, which are chosen following the prescription given in such reference. To test the effect of such ambiguity, the values of (r_D, a_D) were selected as (1.53, 1.06) fm instead of those of Reference [8], (1.51, 1.22) fm. The results were found to be particularly sensitive to the change in a_D, leading to quite different values of optimized depths for the polarization potentials. However, the new set of depths (Table 1) could provide equivalently good fits to both the elastic scattering angular distributions and the fusion cross sections, satisfying in addition the respective dispersion relations. Therefore, the conclusion is that despite the several conditions imposed on the optimization procedure, there are still ambiguities in the potential.

For the ^{9}Be + ^{28}Si system, the results for the eight energies that have been measured [18, 19] are presented in Figure 2, while the optimized parameter values are given in Table 2. All these energies lie above the Coulomb barrier (B = 8.7 MeV, corresponding to 11.5 MeV in the laboratory frame). For the two highest energies, the calculations with W_{int} = 0 MeV show strong oscillations at large angles, but there are not enough data at these angles to reliably confirm or discard a possible physical presence of such oscillations.

It would be nice to have additional data in the region above 90 degrees for such energies, but it is clear that all data existing so far can be well described with smooth curves, using values of W_{int} above 10 MeV, for instance. A value of 40 MeV was used in Reference [6], where data for seven of the eight energies shown in Figure 2 were analyzed also within the framework of the EOM.

With respect to Reference [6], a relevant difference in the present analysis is in the bare potential, which was the SPP here while a double folding potential was used in [6]. A qualitative comparison of the present Figure 2 with Figure 2 of Reference [6] shows that better fits were achieved in the present work, particularly for the lowest energies. In addition, an angular distribution (for E_{Lab} = 13 MeV) not analyzed in [6] was included in this study. Summarizing, the low energy data for ^{6}He + ^{64}Zn and ^{9}Be + ^{28}Si are insensitive to the value of W_{int} and the respective higher energy data do not show the oscillations obtained in calculations with W_{int} = 0 MeV. The existing data for all energies can be well described if a value of 10 MeV or above is used for W_{int}. Values reported in Table 2 correspond to using W_{int} = 50 MeV.

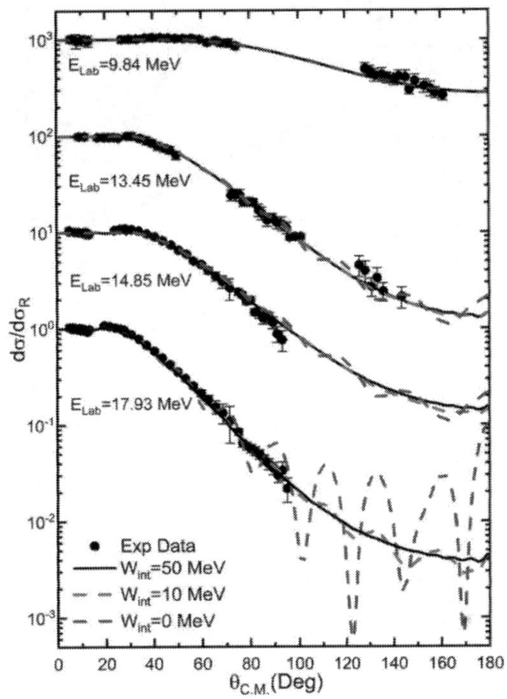

Figure 1. Elastic scattering angular distributions for the ^{6}He + ^{64}Zn system [16, 17] and respective EOM calculations for different values of W_{int}.

The optimized parameters from Tables 1 and 2 satisfy the dispersion relation (3) showed in Figure 3. The error bars are not included because during the optimization process the behavior of some strength parameters was not sensitive to changes in their value, in this sense, further statistical analysis is needed to confirm this behavior.

3 The ^{9}Be + ^{12}C and ^{14}N + ^{9}Be systems

Elastic scattering angular distributions for as many as 11 energies, all above the Coulomb barrier (B = 3.96 MeV), have been measured for the ^{9}Be + ^{12}C system, as shown in Figure 4. The respective parameter values are given in Table 3. In contrast to the two previous cases, it is clear that the data for this system are better reproduced if a value W_{int} = 0 MeV is used in the calculations (r_0, a_0 = 1.0, 0.2 fm were used as geometric parameters).

Therefore, it is apparent that for this light system the total potential is transparent enough so low partial waves survive at close distances, getting reflected by the respective centrifugal barriers, thus producing actual oscillations in the corresponding angular distribution.

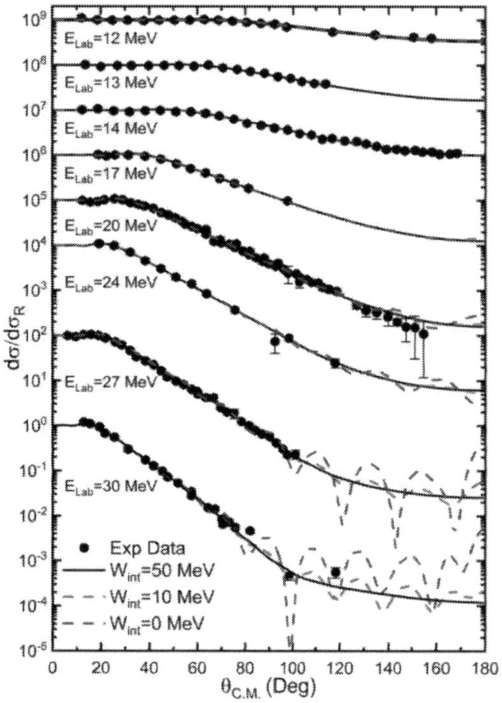

Figure 2. Elastic scattering angular distributions for the ^{9}Be + ^{28}Si system [18, 19] and respective EOM calculations for different values of W_{int}.

Making use of the semi classical deflection function [20], the estimated scattering angle at which each partial wave is reflected has been calculated and indicated with dashed vertical lines in Figure 4. One can see a fairly good correlation between the bumps obtained in the W_{int} = 0 MeV calculation and these scattering angles for the different l values.

Table 3. Optimized parameter values for ^{9}Be + ^{12}C. The first five columns are in MeV. The reduced radius for both polarization potential were r_{0D} = 1.55 fm and r_{0F} = 1.4 fm.

$E_{C.M.}$	V_{0D}	W_{0D}	V_{0F}	W_{0F}	$a_{(D,F)}$ (fm)	χ^2	σ_F (mb)
5.14	0.55	0.85	4.1	5.3	0.50, 0.43	1.367	340.00
6.43	0.20	0.99	4.0	6.5	0.50, 0.43	1.261	481.33
7.43	0.38	0.68	4.0	6.5	0.60, 0.43	2.967	651.64
7.71	0.12	0.60	6.4	8.7	0.60, 0.40	13.782	684.00
8.0	0.16	0.53	4.5	7.3	0.60, 0.40	2.339	719.71
8.29	0.35	0.42	3.63	9.0	0.68, 0.35	6.395	753.25
9.0	0.17	0.25	6.8	9.0	0.82, 0.29	19.187	822.63
9.89	0.21	0.31	4.5	9.3	0.82, 0.29	10.465	839.50
10.86	0.33	0.38	4.7	11.5	0.85, 0.29	18.254	862.32
11.43	0.27	0.35	2.8	10.7	0.85, 0.29	33.924	883.22
12.0	0.35	0.36	3.1	10.5	0.85, 0.26	29.706	890.86

A search for another light system which might show similar behavior as ^{9}Be + ^{12}C, led to the analysis of the ^{14}N + ^{9}Be system, whose Coulomb barrier is at 4.58 MeV. Unfortunately, elastic scattering measurements for only one energy ($E_{C.M.}$ = 9.78 MeV, much higher than the barrier) have been reported for this system [25], with the shortcoming that no dispersion relation can be studied for this case. However, data from the fusion cross section for ^{9}Be + ^{14}N are available [26] to further restrict optimization.

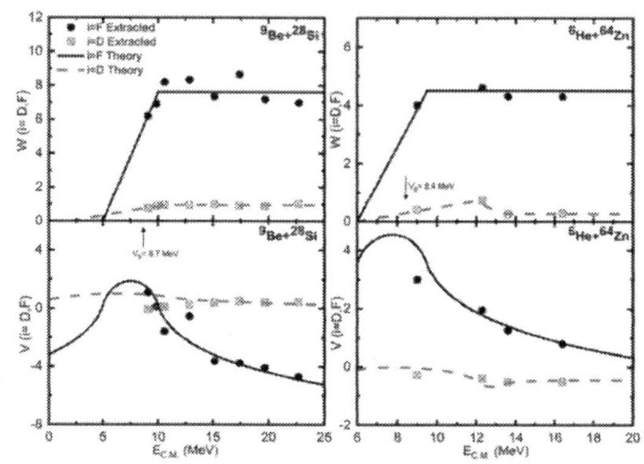

Figure 3. Dispersion relations for the strength parameters of Tables 1 and 2.

The set of potential parameters obtained for ^{14}N + ^{9}Be is given in Table 4, while the results are displayed in Figure 5. The geometric parameters for the direct polarization potential (r_{0D}, a_{0D}) were included in the optimization process due to the limited amount of experimental data available to apply the methodology mentioned in Reference [8]. Similar to ^{9}Be + ^{12}C, the data for this system also present oscillations in the angular distribution, particularly at medium and large angles.

Table 4. Optimized parameter values for ^{14}N + ^{9}Be. The first four columns are in MeV. The geometric parameters for the fusion polarization potential were (r_{0F}, a_{0F}) = (1.40, 0.43) fm.

V_{0D}	W_{0D}	V_{0F}	W_{0F}	r_{0D}, a_{0D} (fm)	σ_F (mb)
-0.01	0.19	-3.60	6.30	1.63, 0.90	762.81

Such oscillations can be reproduced fairly well by using a value of W_{int} = 4 MeV (whose geometrical parameters were r_0, a_0 = 1.0, 0.2 fm) in the EOM calculations. In addition, a nice correlation exists between the bumps in the respective curve and the semi-classical reflection angles corresponding to the different partial waves.

Thus, the two lightest systems analyzed have the common feature that the respective data show distinctive oscillations and the present calculations provide evidence that such oscillations might be due to partial wave reflections facilitated by an unusually weak internal absorption.

4 Fusion Cross Sections

The fusion cross section was obtained as the expected value of the sum of the imaginary absorption potentials using the distorted wave function (χ^+). The latter satisfies

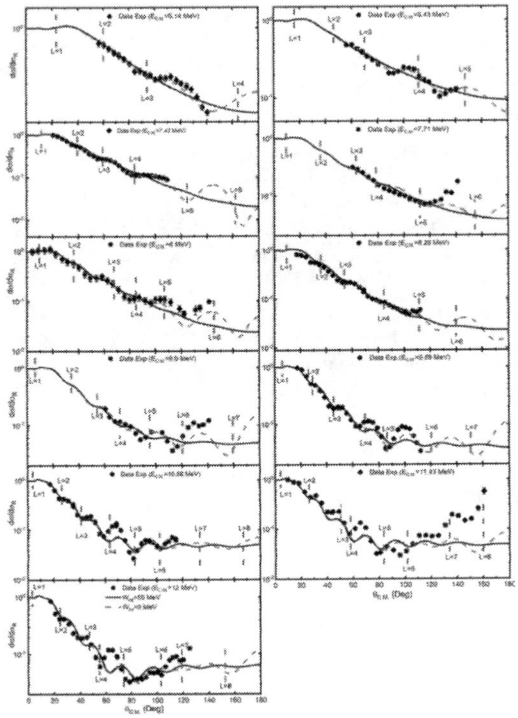

Figure 4. Elastic scattering angular distributions for the ^{9}Be + ^{12}C system [21–24] and respective EOM calculations. Semi-classical estimations of reflection angles for different partial waves are also indicated.

the full Schödinger equation using the effective potential (1). The corresponding expression is:

$$\sigma_F = \frac{2}{\hbar v} < \chi^+ | W_F + W_{int} | \chi^+ >, \tag{4}$$

where v is the relative velocity between both nuclei. The latter procedure was performed by FRESCO.

The results of equation (4) are shown in Figure 6. The fusion cross sections are well reproduced at the 24 energies, considering either the experimental results or the estimates made using Wong's equation (2). The requirement that the strength parameters must reproduce the experimental fusion cross sections helps us to reduce the possible ambiguities of the EOM.

5 Conclusions

The strength parameters provided by the optimization process are able to reproduce the experimental data of elastic scattering for the 24 experimental data sets at different energies for each one (considering the four systems). At the same time, these strength parameters satisfy the dispersion relation for 3 out of 4 systems. The well-known Threshold Anomaly (TA) phenomenon was observed for both ^{6}He+^{64}Zn and ^{9}Be+^{28}Si systems, where an energy was added to the latter one around its threshold compared to the analysis made by Udagawa (2010).

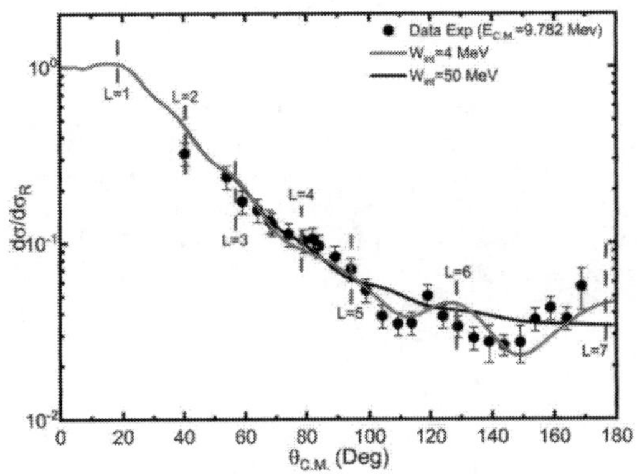

Figure 5. Elastic scattering angular distribution for the ^{14}N + ^{9}Be system [24] and respective EOM calculation. Semi classical estimations of reflection angles for different partial waves are also indicated.

An interesting energy dependence of W_{int} was observed. For all systems, unphysical oscillations appear mainly at large angles, with amplitude depending on the bombarding energy. This behavior can be explained as follows; As the bombarding energy increases, more partial waves are able to overcome the Coulomb barrier and are affected by the W_{int} potential. The amplitude of these oscillations is the result of a competition between the reflection and transmission of partial waves whose angular momenta are lower than certain critical angular momenta related to their strong absorption distance. Finally, the effect of the short range absorption potential W_{int} over the fusion cross section was not significant.

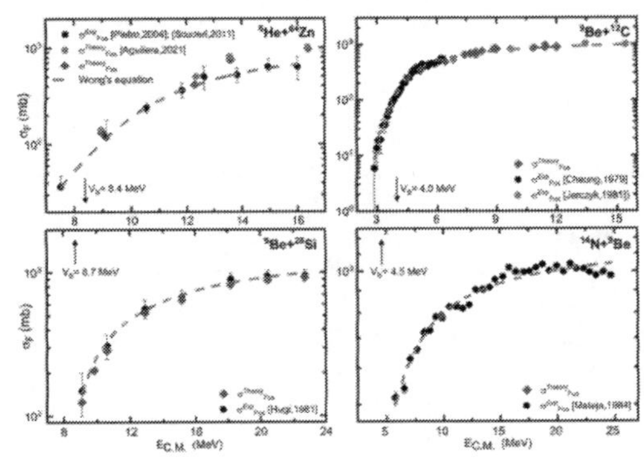

Figure 6. Experimental fusion cross sections an estimates from Wong's equation for compared with the results from the Tables 1-4.

For the two heaviest systems, the oscillatory behavior at large angles that is not present in the experimental data is suppressed theoretically by an increase from 0 MeV to

40-50 MeV in the strength of the W_{int} potential. On the other hand, for the lightest systems, oscillations in the experimental data can be observed for each one of the energies, which are well reproduced by a total or partial suppression in the strength of the W_{int} potential to 4 and 0 MeV respectively.

The semi-classical deflection function has provided valuable information about the partial waves that make up the incident wave function. Those partial waves that have experienced an effective potential with a barrier lower than the center-of-mass energy are mainly affected by the W_{int} potential and show turning points well inside the Coulomb barrier.

References

[1] T. Udagawa, T. Tamura, Heavy-ion subbarrier fusion in terms of the direct reaction technique. Phys. Rev. C **29**, 1922, (1984). https://doi.org/10.1103/physrevc.29.1922

[2] T. Udagawa, *et al.*, Direct reaction description of sub- and above-barrier fusion of heavy ions. Phys. Rev. C **32**, 124, (1985). https://doi.org/10.1103/physrevc.32.124

[3] W. Y. So, *et al.*, Extended optical model analyses of elastic scattering and fusion cross sections for the ^{6}Li + ^{208}Pb system at near-Coulomb-barrier energies using a folding potential,. Phys. Rev. C **75**, 024610, (2007). https://doi.org/10.1103/PhysRevC.75.024610

[4] W. Y. So, *et al.*, Extended optical model analyses of elastic scattering and fusion cross section data for the ^{7}Li + ^{208}Pb system at near-Coulomb-barrier energies using a folding potential. Phys. Rev. C **76**, 024613, (2007). https://doi.org/10.1103/PhysRevC.76.024613

[5] W. Y. So, *et al.*, Extended optical model analyses of elastic scattering and fusion cross section data for the ^{12}C + ^{208}Pb system at near-Coulomb-barrier energies by using a folding potential. Phys. Rev. C **77**, 024609, (2008). https://doi.org/10.1103/PhysRevC.77.024609

[6] W. Y. So, *et al.*, Characteristics of the polarization part of the optical potential for a weakly bound projectile, ^{9}Be. Phys. Rev. C **81**, 047604, (2010). https://doi.org/10.1103/PhysRevC.81.047604

[7] F. Torabi, *et al.*, Systematic study of elastic scattering and fusion induced by weakly bound ^{6}Li on medium mass targets. Threshold anomalies. Nucl. Phys. A **994**, 121661, (2020). https://doi.org/10.1016/j.nuclphysa.2019.121661

[8] E. F. Aguilera, *et al.*, Simultaneous analysis of elastic scattering and fusion in ^{6}He+^{64}Zn: A transition in direct reaction mechanisms, striking threshold anomalies, and halo effects. Phys. Rev. C **104**, 054612, (2021). https://doi.org/10.1103/physrevc.104.054612

[9] F. Torabi, *et al.*, Threshold anomalies in ^{9}Be + ^{12}C. Nucl. Phys. A **1041**, 122793, (2024). https://doi.org/10.1016/j.nuclphysa.2023.122793

[10] C. Y. Wong, Interaction Barrier in Charged-Particle Nuclear Reactions. Phys. Rev. Lett. **31**, 766, (1973). https://doi.org/10.1103/PhysRevLett.31.766

[11] M. A. Nagarajan, *et al.*, Dispersion relation and the low-energy behavior of the heavy-ion optical potential. Phys. Rev. Lett. **54**, 1136, (1985). https://doi.org/10.1103/PhysRevLett.54.1136

[12] G. R. Satchler and W. G. Love, Folding model potentials from realistic interactions for heavy-ion scattering. Phys. Rep **55**, 183, (1979). https://doi.org/10.1016/0370-1573(79)90081-4

[13] N. Keeley, *et al.*Optical model analyses of 6,7Li + 208Pb elastic scattering near the Coulomb barrier. Nucl. Phys. A **571**, 326, (1994). https://doi.org/10.1016/0370-1573(79)90081-4

[14] L. C. Chamon, *et al.*., Toward a global description of the nucleus-nucleus interaction. Phys. Rev. C **66**, 014610, (2002). https://doi.org/10.1103/PhysRevC.66.014610

[15] I. J. Thompson, Coupled reaction channels calculations in nuclear physics. Comp. Phys. Rep. **7**, 167, (1988). https://doi.org/10.1016/0167-7977(88)90005-6

[16] A. Di Pietro, *et al.*., Reactions induced by the halo nucleus ^{6}He at energies around the Coulomb barrier. Phys. Rev. C **69**, 044613, (2004). https://doi.org/10.1103/physrevc.69.044613

[17] J. P. Fernández-García, *et al.*, Breakup mechanisms in the ^{6}He + ^{64}Zn reaction at near-barrier energies. Phys. Rev. C **99**, 054605, (2019). https://doi.org/10.1103/physrevc.99.054605

[18] M. Hugi, *et al.*, Fusion and direct reactions for strongly and weakly bound projectiles. Nucl. Phys. A **368**, 173, (1981). https://doi.org/10.1016/0375-9474(81)90739-9

[19] M. S. Zisman, *et al.*, Dominance of strong absorption in ^{9}Be + ^{28}Si elastic scattering. Phys. Rev. C **21**, 2398 (1980). https://doi.org/10.1103/physrevc.21.2398

[20] R. A. Broglia, *Heavy ion reactions: The elementary processes, parts I&II* (CRC Press, Florida, 2004) 524.

[21] P. H. Barker, *et al.*, One-neutron transfer reactions with beryllium, carbon and oxygen nuclei. Nucl. Phys. A **155**, 401, (1970). https://doi.org/10.1016/0375-9474(70)90903-6

[22] E. Ungricht, *et al.*, Elastic scattering of ^{9}Be on light target nuclei. Nucl. Phys. A **313**, 376, (1979). https://doi.org/10.1016/0375-9474(79)90507-4

[23] L. Jarczyk, *et al.*, Large angle elastic scattering of ^{9}Be ions on carbon isotopes. Nucl. Phys. A **316**, 139, (1979). https://doi.org/10.1016/0375-9474(79)90676-6

[24] R. A. N. Oliveira, *et al.*, Study of ^{9}Be + ^{12}C elastic scattering at energies near the Coulomb barrier. Nucl. Phys. A **856**, 46, (2011). https://doi.org/10.1016/j.nuclphysa.2011.02.005

[25] R. Bock, *et al.*, ^{14}N-induzierte transfer-reaktionen in ^{9}Be von 20 bis 30 MeV. Nucl. Phys. **70**, 481, (1965). https://doi.org/10.1016/0029-5582(65)90449-9

[26] J. F. Mateja, *et al.*, Fusion cross sections for four heavy-ion entrance channels leading to the ^{23}Na compound nucleus. Phys. Rev. C **30**, 134, (1984). https://doi.org/10.1103/physrevc.30.134

Isospin symmetry research at CENS

Xesus Pereira-López[1,*], *Yung Hee* Kim[1], *Sunghan* Bae[1], *Soomi* Cha[1], *Jongwon* Hwang[1], *Dahee* Kim[1], *Chaeyeon* Park[1], and *Laszlo* Stuhl[1]

[1]Center for Exotic Nuclear Studies, Institute for Basic Science, Daejeon 34126, Republic of Korea.

Abstract. In this paper, we outline the research program led by the Center for Exotic Nuclear Studies (CENS) at the Institute for Basic Science aiming to investigate the nuclear structure of proton-rich nuclei and the rich diversity of exotic features they present to further extend our knowledge of the nuclear force. In particular, our interest is focused on improving our understanding of the isospin symmetry, probing the nuclear structure of N~Z nuclei to seek new information on isospin-non-conserving interactions.
The research activities reported in this letter include recent results and details of the experimental approach, discussing recoil-β-tagged experiments using fusion-evaporation reactions as a tool to perform complete spectroscopic studies of proton-rich nuclei in the $f_{7/2}$ and fpg shells. In addition, current efforts to design and develop nuclear instrumentation are also highlighted.

1 Introduction

The similarity of neutron and proton masses and the apparent exchange symmetry between protons and neutrons of the nucleon-nucleon interactions is one of the foundations of nuclear physics. The observed behaviour of nuclei seem to reflect the charge-independence and charge-symmetry of the nuclear interaction led to the definition of the isospin quantum number T by Heisenberg [1] almost a century ago. Since then, the isospin symmetry was widely used in the theoretical description of the nuclear structure. However, if the nuclear interaction was actually the same for protons and neutrons, in the absence of isospin-symmetry-breaking interactions nuclear properties such as masses and excitation energies would depend only on the mass number A and therefore would be identical along an isobaric chain. Naturally, the Coulomb force will lift this degeneracy, although the underlying wave functions are expected to retain their isospin symmetry. Nowadays, there is solid experimental evidence from nucleon-nucleon (NN) scattering data that indicates that isospin symmetry is slightly broken: the neutron-neutron interaction, V_{nn}, is about 1% stronger than the proton-proton interaction, V_{pp}, and the neutron-proton interaction, V_{np}, around 2.5% stronger than the average of V_{nn} and V_{pp} [2]. This confirmation of isospin-symmetry-breaking components in the nuclear interaction ignited a research line focused on improving our understanding of this subtle isospin symmetry violation and study its influence in real nuclear systems, where many-body effects are important.

The isospin symmetry of the nuclear force explains the observation of near-identical structure in members of isobaric multiplets (nuclei with the same mass number A but different total isospin projection $T_z = \frac{N-Z}{2}$), where the excitation energy of states with the same

*e-mail: xpereiralopez@ibs.re.kr

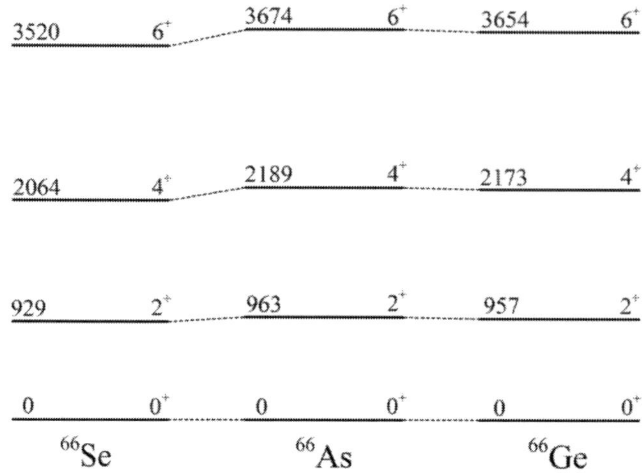

Figure 1. Level schemes of ^{66}Se, ^{66}As and ^{66}Ge show the underlying isospin symmetry between IAS.

isospin number T, namely isobaric analogue states (IAS), is remarkably similar (see figure 1). Energy differences between these IAS arise from isospin-non-conserving interactions, such as the Coulomb interaction. It is common practice to probe the existence of a possible charge-dependent nuclear component by studying Mirror Energy Differences (MED) and Triplet Energy Differences (TED), where differences in excitation energies E_x between isobaric analog states are analyzed in mirror pairs for the former and in all three T_z = -1, 0, 1 members of a T = 1 triplet for the latter:

$$\text{MED}(J^\pi) = E_x(J^\pi, T_z = -T) - E_x(J^\pi, T_z = T) \tag{1}$$

$$\text{TED}(J^\pi) = E_x(J^\pi, T_z = -1) + E_x(J^\pi, T_z = 1) - 2E_x(J^\pi, T_z = 0) \tag{2}$$

However, for A>78 nuclei the proton-rich member of T=1 triplets are so far unknown and therefore comparisons are limited to T_z = 0, 1 members of the triplet, called Coulomb Energy Differences (CED):

$$\text{CED}(J^\pi) = E_x(J^\pi, T_z = 0) - E_x(J^\pi, T_z = 1) \tag{3}$$

CED are more complicated to model [3] since they are sensitive to both charge symmetry (isovector) and charge independence (isotensor), while MEDs and TEDs are governed by isovector and isotensor effects respectively.

The study of these energy differences led to many theoretical ([3–7]) and experimental ([8–22]) investigations devoted to probe the origin of these isospin asymmetries by examining MEDs [8–14], TEDs [15–19] and CEDs [20–22]). These research efforts have shown that electromagnetic effects within the shell model alone cannot explain the observed energy differences (see figure 2), suggesting other effective isospin-non-conserving (INC) interactions are missing from the current models [3–7]. In particular, reference [4] shows that the INC interaction required in the $f_{7/2}$ shell to reproduce experimental data has a very strong J-dependence, for which there is no convincing explanation yet. On the other hand, it suggests a nuclear structure origin and points to a significant missing component in the current models. A comprehensive study of this effect is important to develop nuclear models that successfully account for this phenomenon and are able to reproduce experimental observations.

It is also worth noting that the discussion from [4] was restricted to the $f_{7/2}$ orbital and mostly $\Delta T_z \leq 1$ due to practical reasons, as data available for larger isospin differences or

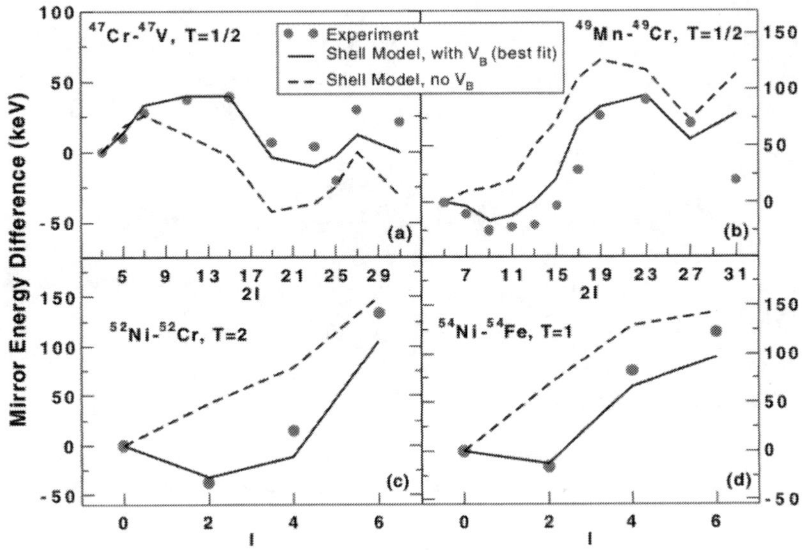

Figure 2. Experimental and predicted MED for four different mirror pairs in the $f_{7/2}$ as a function of angular momentum *l*. Dashed line represents the contribution to the MED of electromagnetic (multipole and monopole) effects, the solid line includes any additional isospin breaking isovector term [5].

beyond Z=28 were scarce due to the difficulty of observing excited states in very proton-rich systems, usually located at the very edge of the proton drip line. However, it is known that several terms contributing to these energy differences increase with ΔT_z, such as the monopole single-particle and radial terms, and therefore it is important study nuclei with large isospin difference to probe the systematics of the INCs contributions and how they scale with ΔT_z. In the same manner, it is important to extract more MED, TED and CED in the fpg shell in order to establish if a isospin-non-conserving (INC) term is required in this shell and whether it also presents a strong dependence on *J*.

1.1 T=0 np pairing

N∼Z nuclei also present the optimal location to probe np pairing effects [20, 23–30]. Pairing correlations have a significant influence in the description of the nuclear structure of N=Z nuclei, where np pairing correlations, in addition to the usual like-nucleon pairing, play a significant role. In neutron-rich nuclei neutrons and protons occupy different single particle orbits, largely suppressing np pairing. Instead, in N∼Z nuclei protons and neutrons are arranged occupying the same orbits, allowing the unusual $T = 0$ np pairing in addition to the normal $T = 1$ (see Figure 3). Despite the experimental efforts, convincing evidence for the former pairing mode still evades observation and hence it remains an open question whether $T = 0$ pairs bear any relevance in the nuclear structure of these nuclei. However, it was recently suggested that a significant delay in the alignment frequency of $g_{9/2}$ pairs of the yrast band observed in ^{72}Kr, ^{76}Sr, ^{80}Zr and ^{88}Ru was a signature of $T = 0$ np pairing, however it is not deemed as conclusive due to possible influence of shape coexistence. Similar interpretations have been reported for N=Z+1 nuclei ^{73}Kr and ^{87}Tc. Of particular interest is the report that suggests spin-aligned np pairs dominate the wave function of the yrast sequence in ^{92}Pd. Subsequent theoretical studies indicate possible contributions from spin-aligned np pairs in the wave functions of the $T = 1$ states in other N=Z $A \sim 90$ odd-odd N=Z nuclei. These studies also indicate the potential for significant contributions from spin-aligned np pairs in the wave functions of the $T = 1$ states in odd-odd N=Z nuclei, such as ^{94}Ag. Therefore, spectroscopic studies of N=Z nuclei in this mass region can provide valuable information to discuss the importance of the $T = 0$ spin-aligned np pairing and its influence in the nuclear structure.

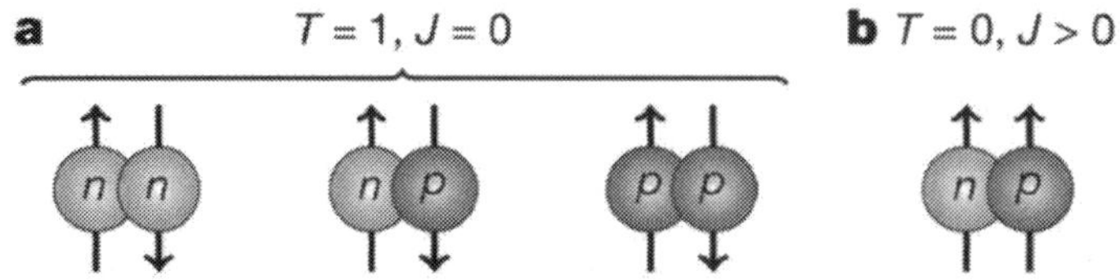

Figure 3. Schematic illustration of the different pairing modes: a) isovector T=1 triplet and b) isoscalar T=0 spin-aligned np pairing [29].

2 Experimental techniques

The perfect testing ground for isospin-symmetry studies are proton-rich nuclei, usually located far from the valley of stability and at the very edge of the proton dripline, being all $Z > N$ nuclei already unbound beyond $Z = 50$. This fact represents the main challenge to progress in our understanding of the isospin symmetry of the nuclear force: the difficulty to explore the most exotic regions of the nuclide chart, at very edge of nuclear existence, and probe the structure of those nuclei with the largest proton excess where evidence of isospin-symmetry-breaking effects can be observed.

Typically, two experimental approaches can generally be considered to study these nuclei: knockout reactions from relativistic radioactive beams and fusion-evaporation reactions with stable beams near the Coulomb-barrier. The limitation in the direct reaction approach comes from the dramatic decrease in the beam intensity of radioactive ion beams as we move closer to the dripline. Unfortunately, a good resolution in excitation energy is important for MEDs, TEDs and CEDs, as energy differences between analog states have been observed to be as low as 2 keV [20]. Therefore, the low beam rates cannot be compensated with the use of thicker targets, as this in turn would compromise the excitation energy resolution due to a bigger uncertainty in the location of the reaction vertex that leads to a worse Doppler correction of the observed γ-rays (already a challenge due to the relativistic beam velocities) and a limited resolution in the energy of the charged particles due to inaccurate corrections for the energy loss on the target.

On the other hand, with fusion-evaporation reactions the biggest challenge comes from the low production cross section of the proton-rich nuclei of interest, often the order of ~200 nb. Hence the challenge here is a clean selection of the reaction channel of interest from the overwhelming background from competing channels, sometimes million times more intense.

2.1 Recoil-β-tagging technique

In recent years, the most notable method to overcome the low production rates has been the recoil-decay-tagging technique. This technique takes advantage of a recoil separator to separate the primary beam from the fusion-evaporation products, using the mass-over-charge ratio and mass slits for further selection. Selected A/q nuclei are then transported to the focal plane detector system, where they are implanted in double-sided silicon strip detector (DSSD). Finally, their subsequent decay can be observed, allowing the identification of the nucleus of interest based on their decay.

This method relies on the observation of a characteristic decay that enables identification: this can be proton, α, β-delayed protons [31–33] or even isomeric γ-rays [34, 35]. In this particular case, however, β decays are used as a tag. Despite β decays being a 3-body process with a continuous energy spectrum that hinders the identification of the decaying nucleus, the recoil-β-tagging technique (RBT), already proven to be successful a number of times [16–21], makes use of the characteristic properties of superallowed β decays: short half-lives and high end-point energies, by requesting the observation of a high-energy β shortly after implantation.

2.1.1 Selected case study: in-beam spectroscopy of ^{94}Ag

The fusion evaporation reaction ^{40}Ca(^{58}Ni,p3n)^{94}Ag has been used to study the odd-odd N=Z nucleus ^{94}Ag at Jyväskylä Accelerator Lab using MARA recoil separator and JUROGAM3 HPGe detector array. The recoil-β-tagging technique has enabled the observation of γ-rays transitions in ^{94}Ag for the first time [21] (see Figure 4). It represents the heaviest odd-odd N=Z nucleus for which excited states are observed, and A=94 (tentative) CEDs are the first in the $g_{9/2}$ orbital and therefore, also the heaviest ever extracted, allowing a major step forward in our understanding of the isospin symmetry breaking interactions.

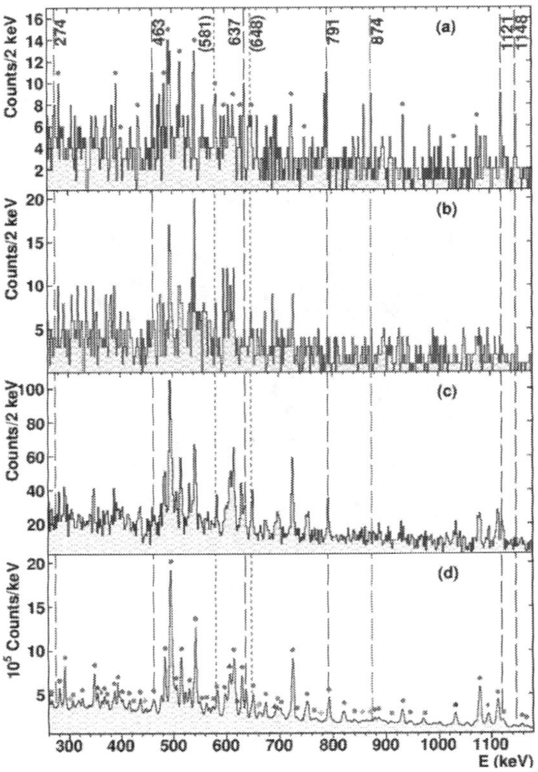

Figure 4. Prompt in-beam γ-ray transitions in ^{94}Ag were selected by requiring that they were recorded in coincidence with one or zero charged-particles (consistent with the p3n reaction channel) and correlated with a short-lived $A = 94$ recoil, observed decaying through a high-energy β within 60 ms from implantation. The resulting γ-ray spectra is shown in panel (a). Panel (b) shows same spectrum as (a) but for longer decay times: between 120 and 180 ms from implantation (other conditions remain the same). Events with 2 or more charged-particles were vetoed in (a), while panel (c) shows same spectrum as (a) but with 2 or more charged-particles were observed in the JYTube (other conditions remain the same). The prompt γ ray spectrum for all $A = 94$ recoils is shown in (d). The ^{94}Ag transitions identified in this work are highlighted in red dashed lines. Symbols in (d) indicate transitions of known contaminants: ^{94}Ru (red circles), ^{94}Rh (green squares), ^{94}Tc (blue triangles), ^{94}Mo (pink diamonds) and ^{90}Mo (orange crosses) - see text for details. Adapted from [21].

3 Nuclear instrumentation

CENS research activities place special emphasis on strengthening the scientific impact of the research performed at RAON (Rare isotope Accelerator complex for ON-line experiments), the new korean accelerator facility. As such, CENS is devoting significant efforts to design, develop and commission state-of-the-art nuclear instruments for operation at RAON [36], including STARK and STARK Jr silicon telescope arrays [37, 38], AToM-X active target, CryoSTAR cryogenic gas cell target, ASGARD HPGe detector array and KoBRA Wien filter [39].

3.1 β-tagging detector

A new position-sensitive plastic scintillator detector has been constructed at CENS for recoil-β-tagging experiments, with a compact design that enables its integration with the focal plane detector system of mass spectrometers at different accelerator facilities worldwide.

This β-tagging detector consists of two orthogonal layers of 9 scintillator bars, each bar being 8 mm wide, therefore the detector presents an active area 7.2 cm long and 7.2 cm wide. The scintillator bars in the first layer are 0.4 cm thick, while the thickness of the bars in the second layer is 2.2 cm, bringing the total thickness of the detector to 2.6 cm, corresponding to approximately to the CSDA range of a 5 MeV β particle [40] (although the geometry of the experimental setup will mean that detected βs will most often meet the detector at an angle, hence the effective thickness will be larger).

Figure 5 presents a picture of the β-detector and a schematic of its design. The dimension of the scintillator bars are $90 \times 8 \times 4$ mm^3 and $80 \times 8 \times 22$ mm^3 in the first and second layer respectively. In both layers, light is collected on both ends using MPPCs S13360 6×6 mm^2 SiPMs manufactured by Hamamatsu. Additionally, all scintillator bars on the second layer have been frustum shaped at both ends to match the SiPMs window, that were attached using optical grease. The scintillator material is Saint Gobain BC-420, with a light yield 64% that of anthracene, a light attenuation length of 140 cm and a decay time of 1.5 ns [41]. In order to optimize the light collection, each bar has been individually wrapped in teflon tape twice.

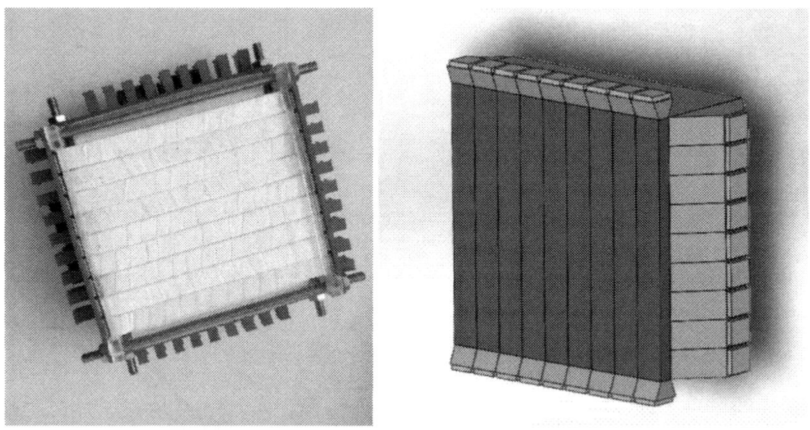

Figure 5. Picture of the position sensitive plastic scintillator custom made at CENS for recoil-β-tagged experiments (left). The schematic illustration of its geometry (right) shows its orthogonal dual-layer design, consisting of scintillator bars with double readout for enhanced position sensitivity.

3.1.1 Detector performance and operation

Energy calibrations are traditionally performed by identifying characteristic and well-understood features of the measured spectra and their associated energy values. In detectors where photoelectric absorption is the dominant interaction mechanism for γ-rays, this is done by associating the location at which clear and well-defined full-energy peaks are observed with the energy value of the corresponding γ-ray transition. However, this simple calibration procedure is not suitable for detectors such as our plastic detector, where Compton scattering is the primary interaction mechanism for γ-rays due to the low Z of the scintillator material. In this scenario, energy calibrations become more challenging, as the recorded spectra with γ-ray sources do not present full-energy peaks suitable for calibration and the continuous energy spectra with low end-point energies of most common β sources also hinder calibration.

Instead, we have used ^{60}Co and ^{137}Cs sources to locate their characteristic Compton edge and take its well-known energy value for calibration purposes. Additional details of this energy calibration method can be found in references [42, 43]. The resulting energy-calibrated spectra for with ^{60}Co and ^{137}Cs are shown in figure 6.

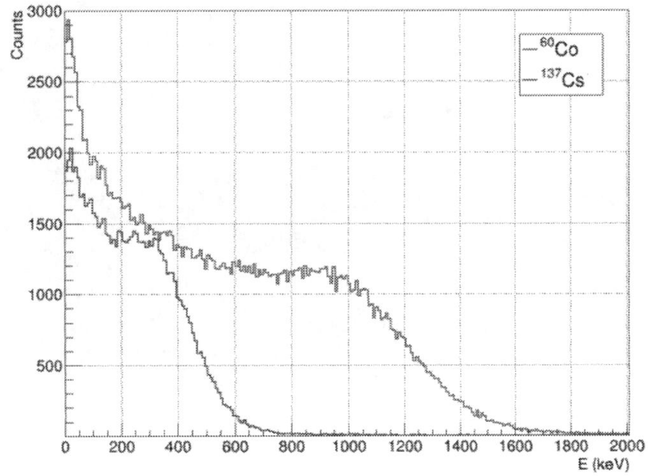

Figure 6. Energy calibrated spectra observed with CENS new plastic scintillator for ^{60}Co (red) and ^{137}Cs (blue) sources, where the characteristic Compton edges are clearly noticeable.

During this procedure, it was noticed that a hit would induce a signal in adjacent bars. In an attempt to quantify this effect, we have studied the relative amplitude of the signals induced in other bars when a hit is observed in the central crystal. Figure 7 shows that an induced signal is observed to some extent in all scintillator bars when a hit is recorded in the central strip, going from nearly 30% in immediately adjacent bars (29.7% and 28.8% in left and right neighbors respectively) to ~10% in the most distant bars (12.4% and 9.6% in bars furthest to the left and right respectively), with its amplitude appearing to decrease linearly with the distance to the hit. Similar behavior was observed in both layers.

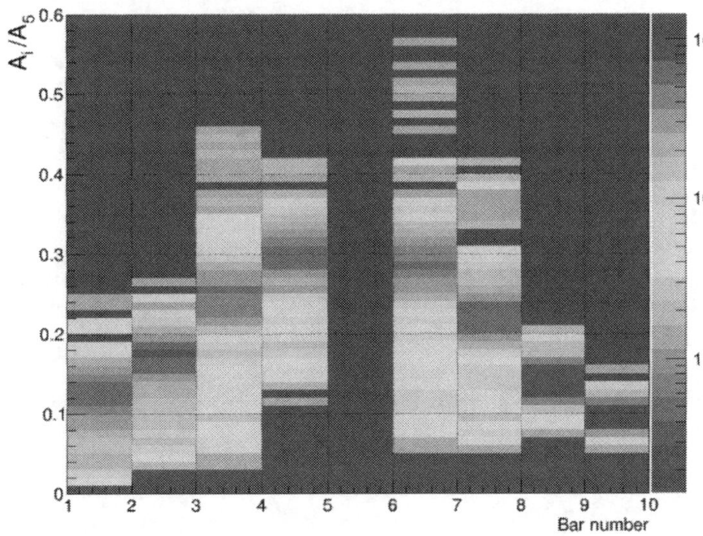

Figure 7. Relative amplitudes of the induced signals in other scintillator bars when a hit is measured in the central one with a ^{60}Co γ-ray source, showing that induced signals in adjacent bars amounts to nearly 30% in adjacent bars.

This phenomenon is arguably due to scattered γ-rays interacting in a second scintillator bar, but crosstalk between adjacent bars, owing to small imperfections in the optical insulation of the individual scintillators, could also contribute to this effect. In order to estimate the light leakage, a similar measurement was performed using a ^{90}Sr β source in order to minimize the induced signals due to scattering. Figure 8 presents the outcome of this study, where we can see that in this situation the induced signal in adjacent bars is often below 10% of the amplitude of the primary signal, dropping to ~5% for more distant crystals. Similar behavior was observed in both layers, while no induced signals where observed between layers.

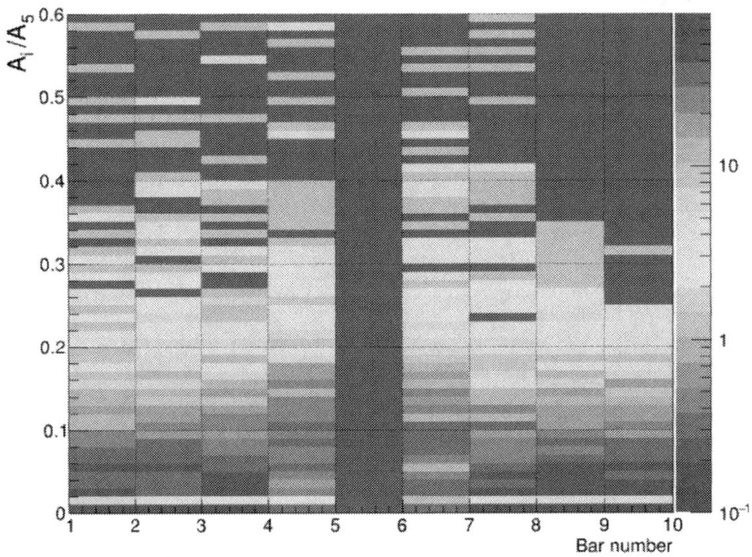

Figure 8. Relative amplitudes of the induced signals in other scintillator bars when a hit is recorded in the central one with a ^{90}Sr β source, showing that the actual crosstalk due to light leakage is below 10%.

4 Conclusion

In summary, we have outlined a research programme conducted at CENS aiming to extend our understanding of the isospin-symmetry-breaking components of the nuclear interaction and investigate its influence in the nuclear structure of N~Z nuclei. This line of research will also address secondary objectives and shed light on other issues of contemporary significance such as the relevance of T=0 spin-aligned np pairing.

The experimental approach to probe the nuclear structure of very proton-rich nuclei and tackle these questions has been discussed and the efforts devoted at CENS to design and develop state-of-the-art instrumentation to pursue this goals have also been presented.

Acknowledgements

The authors want to thank support provided by the Institute for Basic Science (IBS) of the Republic of Korea under grants IBS-R031-D1 and IBS-R031-Y2.

References

[1] W. Heisenberg, Z. Phys. **77**, 1 (1932).
[2] R. Machleidt, Phys. Rev. C **63**, 024001 (2001).
[3] K. Kaneko, Y. Sun, T. Mizusaki, S. Tazaki, Phys. Rev. C **89**, 031302(R) (2014).
[4] M.A. Bentley and S.M. Lenzi, Prog. Part. Nucl. Phys. **59**, 497-561 (2007).
[5] M.A. Bentley, S.M. Lenzi, S.A. Simpson, C.Aa. Diget, Phys. Rev. C **92**, 024310 (2015).
[6] S.M. Lenzi, M.A. Bentley, R. Lau, C.Aa. Diget, Phys. Rev. C **98**, 054322 (2018).
[7] K. Kaneko et al., Phys. Rev. Lett. **109**, 092504 (2012).
[8] J.R. Brown et al., Phys. Rev. C **80**, 011306(R) (2009).
[9] P.J. Davies et al., Phys. Rev. Lett. **111**, 072501 (2013).
[10] S.A. Milne et al., Phys. Rev. C **93**, 024318 (2016).
[11] S.A. Milne et al., Phys. Rev. Lett. **117**, 082502 (2016).
[12] A. Boso et al., Phys. Rev. Lett. **121**, 032502 (2018).
[13] K. Wimmer et al., Phys. Lett. B **785**, 441-446 (2018).
[14] R.D.O. Llewellyn et al., Phys. Lett. B **811**, 135873 (2020).
[15] T.W. Henry et al., Phys. Rev. C **92**, 024315 (2015).
[16] J. Henderson et al., Phys. Rev. C **90**, 051303(R) (2014).
[17] P. Ruotsalainen et al., Phys. Rev. C **88**, 041308(R) (2013).
[18] D.M. Debenham et al., Phys. Rev. C **94**, 004300 (2016).
[19] G. L. Zimba et al., Phys. Rev. Lett. **134**, 022502 (2025).
[20] B.S. Nara Singh et al., Phys. Rev. C **75**, 061301(R) (2007).
[21] X. Pereira-López et al., Eur. Phys. J. A **59**, 44 (2023).
[22] G. de Angelis et al., Eur. Phys. J. A **12**, 51-55 (2001).
[23] G. de Angelis et al., Phys. Lett. B **415**, 217–222 (1997).
[24] P.J. Davies et al., Phys. Rev. C **75**, 011302(R) (2007).
[25] S.M. Fischer et al., Phys. Rev. Lett. **87**, 132501 (2001).
[26] N.S. Kelsall et. al., Phys. Rev. C **64**, 024309 (2001).
[27] B.S. Nara Singh et al., Phys. Rev. C **86**, 041301(R) (2012).
[28] X. Liu et al., Phys. Rev. C **104**, L021302 (2021).
[29] B. Cederwall et al., Nature **469**, 68–71 (2011).
[30] B. Cederwall et al., Phys. Rev. Lett. **124**, 062501 (2020).
[31] E. Paul et al., Phys. Rev. C **51**, 78 (1995). tagging
[32] D.G. Jenkins et al., , Phys. Rev. C **62**, 021302 (2000).
[33] J. Pakarinen et al., Phys. Rev. C **80**, 031303 (2009).
[34] D.M. Cullen et al., Phys. Rev. C **66**, 034308 (2002).
[35] P.J.R. Mason et al., Phys. Rev. C **81**, 024302 (2010).
[36] S. Ahn, D.S. Ahn, S. Kim, et al., J. Korean Phys. Soc. **82**, 537–546 (2023).
[37] D. Kim et al., Nucl. Instrum. Methods Phys. Res. B **540** 30-32 (2023).
[38] X. Pereira-López et al., Nucl. Instrum. Methods Phys. Res. B **541** 134-136 (2023).
[39] J.W. Hwang et al., Nucl. Inst. Meth. Phys. Res. B **541**, 1-3 (2023).
[40] M.J. Berger, J.S. Coursey, M.A. Zucker, J. Chang, NIST standard reference database **124** (2017), http://dx.doi.org/10.18434/T4NC7P.
[41] Saint-Gobain Ceramics & Plastics Inc., BC418, BC420, BC422 Premium Plastic Scintillators Data Sheet.
[42] E.R. Siciliano et al., Nucl. Instrum. Methods Phys. Res. A **594** 232–243 (2008).
[43] H. Joukainen et al., Nucl. Instrum. Methods Phys. Res. A **1027** 166253 (2022).

EPJ Web of Conferences **333**, 01009 (2025) https://doi.org/10.1051/epjconf/202533301009
XLVI Symposium on Nuclear Physics 2025

Cluster structure of the ground state of light exotic nuclei beyond α clustering

Tania Zanatta-Martinez[1,*], *Valerian* Girard-Alcindor[1,**], and *Didier* Beaumel[1,***]

[1]IJCLab, Laboratoire De Physique Des 2 Infinite Irène Joliot-Curie, 15 Rue Georges Clemenceau, 91400 Orsay

Abstract. Preliminary results of a study on triton clustering in neutron-rich ^{10}Be and ^{12}Be isotopes through triton (p,α) and alpha (d,^{6}Li) transfer are presented. The experiment was performed using the LISE fragmentation beam line of GANIL, making use of the MUGAST-EXOGAM-ZDD setup, which ensures accurate measurement of charged particles and gamma-rays. The experiment, currently under analysis, aims to compare the measured differential cross sections with DWBA calculations performed using microscopic and cluster wave functions derived from models such as AMD or THSR. This ongoing analysis aims to provide quantitative insights into triton clustering in neutron-rich beryllium isotopes. The detection setup used in this experiment is presented, along with preliminary results on the data analysis.

1 Introduction

Clustering is a universal phenomenon in composite systems occurring at all scales of the material world. At bigger scales, such clustering phenomenon may exist in giant objects in the universe like solar systems, galaxy cumulus, or, in a more particular example, the crust of neutron stars in core collapse supernovae. In smaller systems, we can observe clustering within the atomic nucleus as a composition of nucleons. Within the nucleus, correlations among nucleons play a decisive role in understanding the properties of the whole atomic nuclei and, more broadly, nuclear matter. Within the latter, nucleons are expected to form light nuclear clusters, with the most expected being the α particle. It is a particular example of a compact 4 nucleon correlation due to its strong binding energy, beneficial for its cluster formation. Theories have predicted α clustering for light self-conjugate nuclei such as ^{8}Be, ^{12}C and ^{16}O, pictured by the so-called Ikeda diagram [1]. Following these predictions, microscopic theories on α clustering have been developed presuming strong correlations between nucleons in clusters and weak inter-cluster correlations.

1.1 Clustering in Neutron-Rich Nuclei

Experimental evidence of cluster structures has been found for light nuclei in their and excited states [2] at the vicinity of the cluster emission thresholds as shown by Ikeda [1]. Nevertheless, closeness to particle emission thresholds doesn't seem to be the only explanation for for

*e-mail: zanatta-martinez@ijclab.in2p3.fr
**e-mail: girard-alcindor@ijclab.in2p3.fr
***e-mail: beaumel@ijclab.in2p3.fr

clustering. For example, there are strong indications of alpha clustering in the g.s. of light neutron-rich nuclei such as Be and B isotopes.

Predictions from generalized density functional theories in infinite nuclear matter, in which light clusters are included as explicit degrees of freedom, have been made as shown in figure 1 (left). These calculations, at total proton fraction $Y_P = 0.4$, predict that the light clusters d, t, ^{3}He and ^{4}He to reach a maximum saturation at approximately 1/10 of the nuclear saturation density. This model predicts the α clustering at a similar magnitude compared to other light clusters. On the right, we can see the fraction of light clusters as a function of the total baryon density in nuclear matter. In particular, (a) and (c) show that for neutron-rich nuclear matter, triton clustering is predicted to be enhanced and alpha clustering to be reduced.

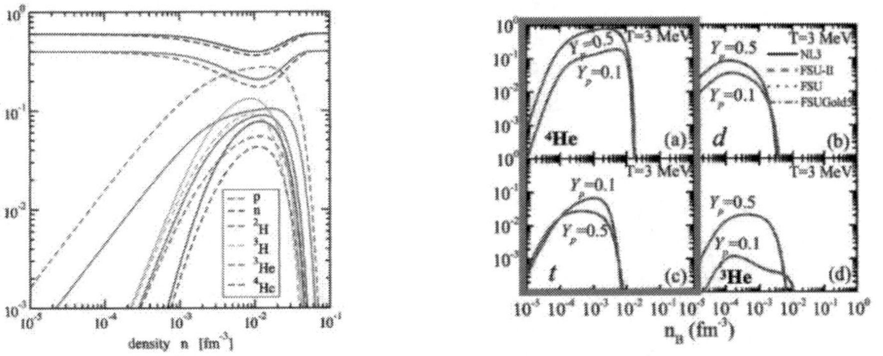

Figure 1. Left: Predictions from generalized density functional theories by S. Typel and by Z. W. Zhang in infinite nuclear matter [6]. Right: Fraction of light clusters as a function of the total baryon density in nuclear matter [10].

Furthermore, studies on medium-to-heavy nuclei have been conducted at the RIKEN and RCNP facilities over the last decade. In medium-heavy nuclei, an experiment by Tanaka et al. (2021) [3] provided experimental evidence of the behavior of α clusters on the surface of the neutron-rich $^{112-124}$Sn isotope chain. This experiment, using the α-knockout reaction $(p,p\alpha)$, demonstrated a decrease in α clustering as the isotopes became more neutron-rich. The paper explores the competition with the neutron skin, but the theoreticians who reproduced the experimental results carried out calculations that also suggest alternative interpretations.

1.2 Experimental Studies on Clustering in Light Nuclei

An experiment was conducted at RIKEN by my team at IJCLab to explore the cluster structure of neutron-rich Be isotopes using quasi-free cluster-knockout reactions $(p,p\alpha)$. It aimed to investigate alpha clustering at the surface of neutron-rich beryllium isotopes up to the drip line by measuring the triple-differential cross-sections (TDX) of the ^{10,12,14}Be$(p,p\alpha)$ reactions in inverse kinematics under quasi-free conditions. The TDX for the ^{10}Be$(p,p\alpha)^6$He reaction at 150 MeV/u and were compared to Distorted Wave Impulse Approximation (DWIA) calculations within a microscopic framework incorporating Tohsaki-Horiuchi-Schuck-Ropke (THSR) and Antisymmetrized Molecular Dynamics (AMD) alpha-cluster wave functions. Notably, this AMD approach does not assume preformed clusters; instead, clustering naturally emerges from the mean-field. The remarkable agreement in both shape and magnitude between the measured and calculated TDX provides direct evidence of the molecular-type cluster structure in the ground states of ^{10}Be and ^{12}Be [12].

In addition, this experiment yielded data for the triton knockout reaction ^{14}Be(p,pt)^{11}Li. The extracted TDX appeared comparable to that of the ^{14}Be(p,$p\alpha$)^{10}He reaction. However, due to residue acceptance constraints, data were not obtained for the ^{10,12}Be(p,pt) reactions. This work presents preliminary results on an experiment performed to investigate cluster formation in light nuclei by means of an alternative approach to knockout reactions: cluster transfer reactions.

2 Present study

Cluster transfer reactions offer a valuable alternative to cluster knockout reactions, as both probe the same nuclear structure information, the overlap function between the initial and residual nuclei A and B, where $A = B + X$ (with X being the transferred cluster). This overlap function corresponds to the cluster radial wave-function in nucleus A and plays a crucial role in the transition amplitude of both reactions within the DWBA and DWIA reaction models.

Previous studies [13] have compared the results of alpha knockout and stripping reactions, showing that the target-mass dependence of the ground-state spectroscopic factors extracted from $(p, p\alpha)$ reactions aligns well with those obtained from the alpha-transfer $(^6\text{Li}, d)$ reaction. Therefore, a study of clustering in Be isotopic chain would confirm RIKEN's hint of a strong triton clustering and would allow to study the evolution of triton clustering with neutron number. To directly assess the validity of the knockout/transfer comparison within our data, the angular distribution of the α/^{6}Li will be analyzed by Distorted Wave Born Approximation (DWBA) calculations with microscopic wave functions, such as those from AMD, comparing the (d, ^{6}Li) alpha transfer reaction to the $^{10}Be(p,\ p\alpha)$ reaction. The experiment aims to study triton clustering in neutron-rich beryllium isotopes ^{10,12}Be through the triton pickup (p,α) reaction. At GANIL we have measured the differential cross-section of the ground-state transition for the reactions: ^{10}Be(p, α), ^{12}Be(p, α) at 38 MeV/u and ^{10}Be(d,^{6}Li) at 27 MeV/u.

3 Experimental Setup and Methodology

The experiment was conducted using fragmentation beams, produced by the LISE spectrometer at GANIL in June 2024. The MUGAST array consists of 4 MUST2 telescopes [18], which are made of two layers, a 300 μm 128x128 DSSD and 4x4 CsI crystals, the EXOGAM high purity germanium gamma-ray spectrometer [19] and the new zero degree detection (ZDD) system composed of a drift chamber (DC), ionization chamber and (IC) plastic detector, installed in experimental room D6. The MUGAST array was used to measure the energies and scattering angles of all the emitted particles (^{6,7}Li and ^{4,6}He) to allow the full kinematical reconstruction of the reactions, while the EXOGAM setup was used to infer the contribution of excited states in the lithium residues. The beam energy, position, and timing were monitored on an event-by-event basis using MWPCs called the CATS detectors [20]. One CATS detector operating in timing mode was placed at the D4 position, 2.6 m before a second pair of CATS detectors in tracking mode positioned upstream of the reaction chamber, as shown in Figure 2. Two CH_2 and CD_2 targets, each with a thickness of 5 mg/cm^2, were installed inside the reaction chamber. The time-of-flight measurement between D4 and D6 is expected to provide a beam energy resolution of approximately 1% (FWHM) and a detection efficiency close to 100% for beryllium isotopes at 40 MeV/u.

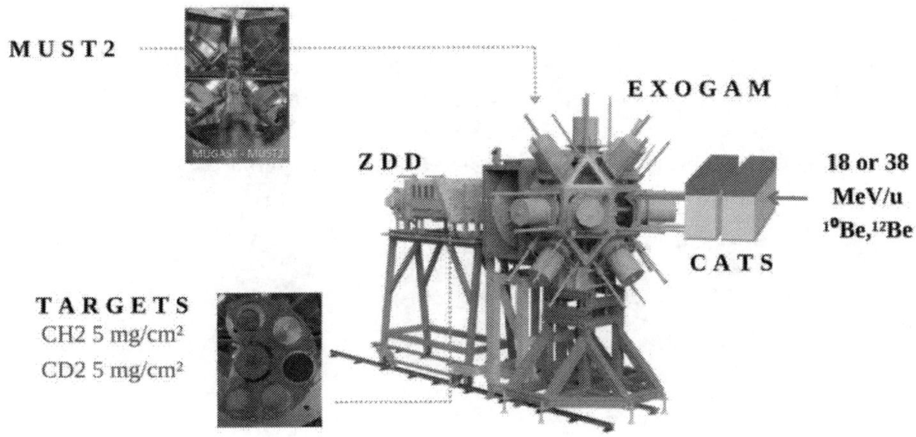

Figure 2. Graphical representation of detection system in D6, LISE beam line @GANIL including CATS, MUST2, ZDD and EXOGAM detectors.

4 Preliminary results

4.1 ^{10}Be$(p, \alpha)^7$Li @ 38 MeV/u

Kinematic lines are shown on the left side of figure 3 for both light and heavy reaction products. In both cases, the solid line represents the calculated kinematic lines from the ^{10}Be$(p, \alpha)^7$Li reaction at the same experimental beam energy of 38 MeV/u.

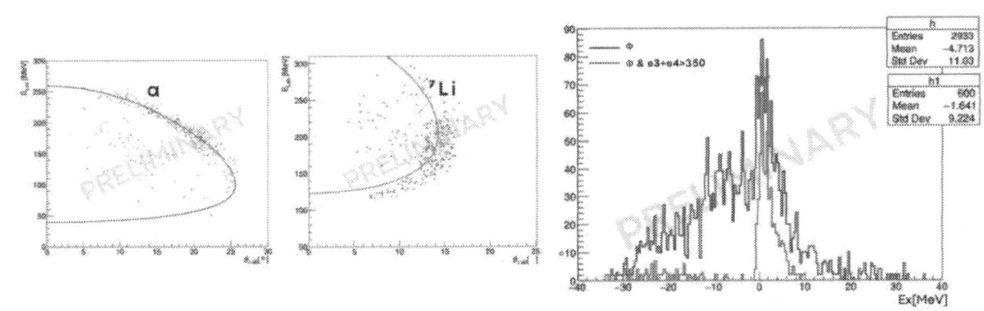

Figure 3. Energy and angle correlations for the ^{10}Be$(p, \alpha)^7$Li reaction. **Left:** Experimental data (dots) compared to calculated kinematic lines (solid). **Right:** Reconstructed excitation energy via invariant mass with (blue line) and without (red line) carbon background (see text).

The dots represent experimental data selected based on particle identification of α and ^{7}Li particles in the MUST2 detector using the $\Delta E - E$ technique. There is an excellent agreement between the event distribution and the simulated kinematic lines. The reconstructed excitation energy of the ground state of ^{10}Be, obtained via the invariant mass method, is shown on the right side of figure 3. The blue line represents the total events, while the carbon background is removed in the red line by adding conditions on the sum of the particles energies. A clear peak at 0 MeV suggests the identification of the mentioned ^{10}Be ground state.

4.2 ^{10}Be$(d,^{6}Li)^{6}$He @ 17 MeV/u

In analogy to the previous case, figure 4 presents the kinematic lines, as well as the reconstructed excitation energy of the ground state of ^{10}Be for the ^{10}Be$(d,^{6}$Li$)^{6}$He reaction. A mismatch between the simulated and experimental kinematic lines suggests that the observed double peak in the excitation energy spectrum may originate from detector calibration issues.

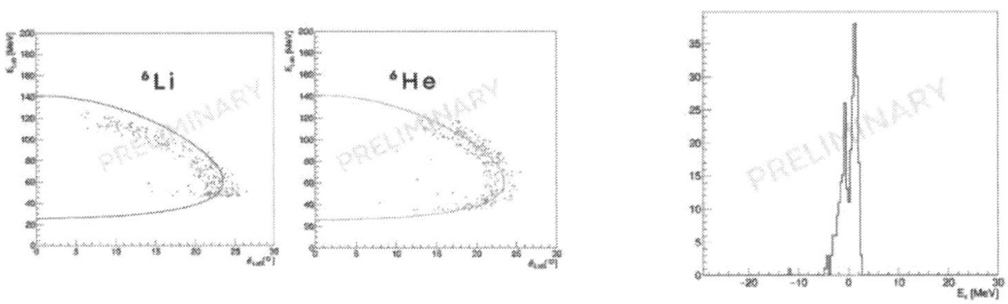

Figure 4. Energy and angle correlations for the ^{10}Be$(d,^{6}Li)^{6}$He reaction. **Left:** Experimental data (dots) compared to calculated kinematic lines (solid). **Right:** Reconstructed excitation energy via invariant mass with (blue line) and without (red line) carbon background.

4.3 ^{12}Be$(p,\alpha)^{9}$Li @ 38 MeV/u

A similar analysis to the previous cases was applied to reaction ^{12}Be$(p,\alpha)^{9}$Li. In this case, no kinematic lines could be reconstructed, resulting in a non identifiable excitation energy peak.

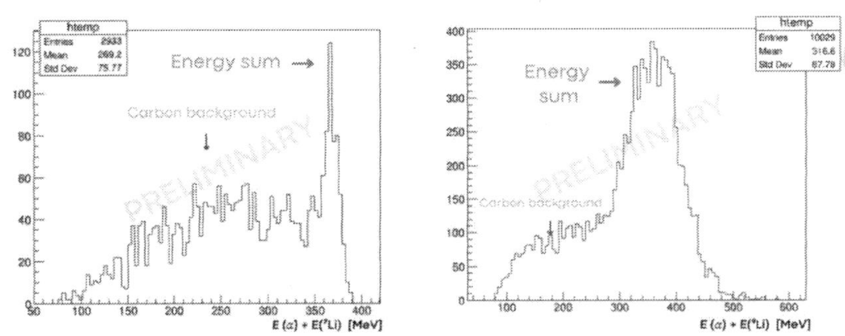

Figure 5. Sum of energies E_{light} + E_{heavy} of light and heavy products. Left: ^{10}Be$(p,\alpha)^{7}$Li. Right: ^{12}Be$(p,\alpha)^{9}$Li

A comparison of the total energy sum E_{light} + E_{heavy} of the detected particles between the reactions ^{10}Be$(p,\alpha)^{7}$Li and ^{12}Be$(p,\alpha)^{9}$Li is shown in figure 5. On the left spectra, the narrow peak from E_{α} + $E_{^{7}Li}$ represents the events correlated by energy from the ^{10}Be$(p,\alpha)^{7}$Li reaction. It can be distinguished from the broad structure attributed to the carbon background. In the spectrum on the right, a similar structure arises from correlated events above the carbon background. However, the broader peak suggests a wider energy distribution compared to the previous case, which may result from a large energy spread of the beam.

5 Conclusions and perspectives

The study of clustering phenomena in neutron-rich isotopes remains an exciting and evolving field in nuclear physics. In particular, one open question is whether different clustering modes, such as triton clustering, compete with each other. This proceeding focuses on this latter hypothesis. Although no definitive conclusion can be drawn at this stage, preliminary results suggest, at the very least, that the triton transfer cross-section is larger for ^{12}Be than for ^{10}Be. To reach a conclusion, further analysis needs to be performed. This will require improved resolution through beam tracking, the determination of absolute cross-sections, and a comparison between knockout and transfer data using the same AMD wave function. The results will provide insight for further applications to other isotopic chains.

References

[1] K. Ikeda, N. Takigawa, and H. Horiuchi, Prog. Theor. Phys. Suppl. **E68**, 464 (1968).
[2] M. Freer, H. Horiuchi, Y. Kanada-En'yo, Ulf-G. Meißner, Rev. Mod. Phys. **90**, 035004 (2018).
[3] J. Tanaka et al., Science **371**, 260-264 (2021).
[4] G. Röpke, Phys. Rev. C **79**, 014002 (2009).
[5] S. Typel et al., Phys. Rev. C **81**, 015803 (2010).
[6] S. Typel, J. Phys. Conf. Ser. **420**, 012078 (2013); S. Typel, H. H. Wolter, G. Röpke, and D. Blaschke, Eur. Phys. J. A **50**, 17 (2014).
[7] K. Hagel, J. B. Natowitz, and G. Röpke, Eur. Phys. J. A **50**, 39 (2014).
[8] G. Giuliani, H. Zheng, and A. Bonasera, Prog. Part. Nucl. Phys. **76**, 116 (2014).
[9] T. Uesaka and J. Zenihiro, The ONOKORO Project.
[10] Z.-W. Zhang and L.-W. Chen, Phys. Rev. C **95**, 064330 (2017).
[11] S. Typel, Phys. Rev. C **89**, 064321 (2014).
[12] P. J. Li et al., Phys. Rev. Lett. **131**, 212501 (2023). doi:10.1103/PhysRevLett.131.212501.
[13] T. A. Carey et al., Phys. Rev. C **29**, 1273 (1984).
[14] F. Pellegrini et al., Phys. Rev. C **28**, 42 (1983).
[15] F. Brunner et al., J. Phys. G: Nucl. Phys. **6**, 473 (1980).
[16] G. Gambarini et al., Nucl. Phys. A **126**, 562 (1969).
[17] T. Yamaya et al., Phys. Lett. B **90**, 219 (1980).
[18] E. Pollacco et al., Eur. Phys. J. A **25** (Suppl 1), 287–288 (2005).
[19] J. Simpson et al., Acta Phys. Hung. New Ser. Heavy Ion **11**, 159–188 (2000). in2p3-00438782.
[20] S. Ottini-Hustache et al., Nucl. Instrum. Methods Phys. Res. A **431**, 476–484 (1999).

EPJ Web of Conferences **333**, 02001 (2025) https://doi.org/10.1051/epjconf/202533302001
XLVI Symposium on Nuclear Physics 2025

Nuclear Structure Aspects of Ordinary Muon Capture

Osvaldo Civitarese[1,2]

[1]Institute of Physics. CONICET (IFLP)
[2]Department of Physics. University of La Plata. Argentina

Abstract. A central problem in today's physics is the determination of neutrino properties, e.g; its mass and Majorana nature. The problem is just in the intersection between nuclear and particle physics, both experimental and theoretical ones. In this talk we shall discuss an alternative to fix the uncertainties attributed to the nuclear structure part of the problem and it consists of the calculation of ordinary muon capture processes (OMC) leading to nuclei which participate in nuclear double beta decay chains.

1 Introduction

Neutrino-less double beta-decay ($0\nu\beta\beta$ decay) [1, 2] is both one of today's major experimental challenges and, at the same time, the most promising means of observing lepton-number violation. It may provide evidences that neutrinos are their own antiparticles (Majorana type), and give information on the absolute effective neutrino mass, right-handed leptonic current, left-right coupling constants, and also an insight in the matter-antimatter asymmetry of the Universe. In some large scale experimental efforts nuclei which are candidates to decay via double beta decay, are measured using ordinary muon capture (OMC) as a tool. In this note I shall address the nuclear structure issues involved in the theoretical description of OMC.

2 Basic notions about ordinary muon capture(OMC)

OMC rates are calculated by using the formalism of Morita and Fujii [3, 4], with exact muon (μ) relativistic (Dirac) wave functions. The OMC rate is given by the expression

$$W = 2P(2J_f + 1)\left(1 - \frac{q}{m_\mu + AM}\right)q^2$$

where the momentum exchange q is written as

$$q = (m_\mu - W_0)\left(1 - \frac{m_\mu}{2(m_\mu + AM)}\right).$$

J_f is the angular momentum of the final-state , M the average nucleon mass and m_μ the rest mass of the muon, $W_0 = M_f - M_i + m_e + E_X$ contains the masses of the initial and final nuclei and E_X denotes the excitation energy in the final nucleus. For the OMC on ^{136}Ba it will be ^{136}Cs.

The rate function P contains the nuclear matrix elements, phase-space factors and combinations of the weak axial-vector coupling $g_{\rm A}$ and induced pseudo-scalar coupling $g_{\rm P}$[1, 2] The effect of the two body currents is taken into account by the effective coupling

$$g_{\rm A} \to (1 + \delta_a(q^2))g_{\rm A}$$

and

$$g_{\rm P} \to \left(1 - \frac{q^2 + m_\pi^2}{q^2}\delta_a^{\rm P}(q^2)\right)g_{\rm P}$$

where the tbc contributions $\delta_a(q^2)$ and $\delta_a^{\rm P}(q^2)$ are approximated by the normal-ordered one-body part of the chiral two-body currents. In order to calculate OMC observables, for the case of a heavy mass nuclei like ^{136}Ba $\to$ ^{136}Cs, we shall apply two nuclear many body techniques like the quasiparticle random phase approximation (QRPA) and its symmetry restoring version, the particle number projected quasiparticle random phase approximation (PNQRPA), which is meant to restore the particle number symmetry broken by the use of the BCS transformation [5, 6]. The values of the occupation amplitudes v_j and vacancy amplitudes u_j, of the BCS transformations, are derived by minimizing the expectation value

$$\begin{aligned} E_0^N &= I_E/I_0 \\ I_E &= \langle \mathrm{BCS}|\hat{H}\hat{P}_N|\mathrm{BCS}\rangle \\ I_0 &= \langle \mathrm{BCS}|\hat{P}_N|\mathrm{BCS}\rangle \end{aligned}$$

of the nuclear Hamiltonian $\hat{H}$ in the number-projected vacuum

$$\begin{aligned} &\langle \mathrm{PBCS}|\hat{H}|\mathrm{PBCS}\rangle/\langle \mathrm{PBCS}|\mathrm{PBCS}\rangle \\ &|\mathrm{PBCS}\rangle = \hat{P}_N|\mathrm{BCS}\rangle \\ &\hat{P}_N = \frac{1}{2\pi}\int_0^{2\pi} e^{i(N-\hat{N})\phi}d\phi \end{aligned}$$

with $\hat{N}$ being the particle-number operator. The variation $\nabla(v_i, u_i)$ with respect to v_j and u_j leads to the variational equation

$$I_0\nabla(v_i, u_i)I_E - I_E\nabla(v_i, u_i)I_0 = 0$$

which leads to the determination of the occupation probabilities $v_i^2 = 1 - u_i^2$, and thus the structure of the projected BCS vacuum $|\mathrm{PBCS}\rangle$. The PNQRPA equations of motion look like the ordinary QRPA equations of motion when replacing the A and B matrices by their particle-number projected counterparts $A(N \mp 1, Z \pm 1)$ and $B(N \mp 1, Z \pm 1)$:

$$\begin{pmatrix} A(N \mp 1, Z \pm 1) & B(N \mp 1, Z \pm 1) \\ -B(N \mp 1, Z \pm 1) & -A(N \mp 1, Z \pm 1) \end{pmatrix}\begin{pmatrix} X \\ Y \end{pmatrix} = \Omega\begin{pmatrix} X \\ Y \end{pmatrix}$$

where Ω are the PNQRPA energies. The PNQRPA matrices are defined as

$$\begin{aligned} &A_{\alpha\beta}(N \mp 1, Z \pm 1) \\ &= \langle \mathrm{PBCS}|b_\alpha(J^\pi, M)\hat{P}_{N\mp1}\hat{P}_{Z\pm1}[\hat{H}, b_\beta^\dagger(J^\pi, M)]|\mathrm{PBCS}\rangle \\ &B_{\alpha\beta}(N \mp 1, Z \pm 1) \\ &= \langle \mathrm{PBCS}|b_\alpha(J^\pi, M)\hat{P}_{N\mp1}\hat{P}_{Z\pm1}[\hat{H}, \tilde{b}_\beta(J^\pi, M)]|\mathrm{PBCS}\rangle \end{aligned}$$

where the angular-momentum-coupled proton-neutron quasiparticle pairs are defined as

$$
\begin{aligned}
b^{\dagger}_{\alpha}(J^{\pi}, M) &= \sum_{pn} \eta^{\alpha}_{pn} A^{\dagger}(pn, J^{\pi}, M) \\
\tilde{b}_{\alpha}(J^{\pi}, M) &= \sum_{pn} \eta^{\alpha}_{pn} \tilde{A}(pn, J^{\pi}, M)
\end{aligned}
$$

The above-described formalism leads to states of an odd-odd nucleus $(N-1, Z+1)$ or $(N+1, Z-1)$ by using the phonon-creation operator $\Gamma^{\dagger}$:

$$
\begin{aligned}
&|N \mp 1, Z \pm 1, kJ^{\pi}M\rangle \\
&= \Gamma^{\dagger}(N \mp 1, Z \pm 1, kJ^{\pi}M)|\mathrm{PNQRPA}\rangle
\end{aligned}
$$

with

$$
\begin{aligned}
&\Gamma^{\dagger}(N \mp 1, Z \pm 1, kJ^{\pi}M) \\
&= \textstyle\sum_{\alpha} \left[X^{N\mp1,Z\pm1}_{\alpha}(J^{\pi}k) b^{\dagger}_{\alpha}(J^{\pi}, M) - Y^{N\mp1,Z\pm1}_{\alpha}(J^{\pi}k)\tilde{b}_{\alpha}(J^{\pi}, M)\right]
\end{aligned}
$$

The PNQRPA-computed one-body transition densities (OBTD) are of the form

$$
\begin{aligned}
&(J^{\pi}, k||[c^{\dagger}_{n}\tilde{c}_{p}]_{J}||0^{+}_{gs}) \\
&= -\left[v_{p}u_{n}\bar{X}_{pn}(J^{\pi}, k) + u_{p}v_{n}\bar{Y}_{pn}(J^{\pi}, k)\right]
\end{aligned}
$$

where

$$
\begin{aligned}
\bar{X}_{pn}(J^{\pi}, k) &= \sum_{\alpha} \eta^{\alpha}_{pn} X^{N\mp1,Z\pm1}_{\alpha}(J^{\pi}k) \\
\bar{Y}_{pn}(J^{\pi}, k) &= \sum_{\alpha} \eta^{\alpha}_{pn} Y^{N\mp1,Z\pm1}_{\alpha}(J^{\pi}k)
\end{aligned}
$$

With these elements concerning the relevant theoretical structure aspects of the problem, we are in conditions to calculate the partial contributions to the decay rates.

3 Discussion and Results

The first aspect to take into account is the dependence of both QRPA and PNQRPA 1^{+} states upon g_{pp}[5]. In Table 1 the calculated energy levels of the double odd mass nuclei ^{136}Cs are shown, for both the QRPA and PNQRPA nuclear model descriptions, for the chosen value of the particle-particle coupling constant (g_{pp}) of the proton-neutron channel of the interaction. With the calculated wave functions and energies of the spectrum of ^{136}Cs we have obtained the OMC transition densities to the lowest $J^{\pi} = 1^{+}, 2^{+}$ states . The results are shown in Table 2 and in Table 3, respectively. Finally, the OMC rates for the low lying states of ^{136}Cs, with energies below 1 MeV, are shown in Tables 4 and 5.

It is noticed, from the results shown in both tables Table 4 and Table 5, that the OMC rates for the PNQRPA case are significantly larger than those resulting from the use of the QRPA method, for the dominant $J^{\pi} = 1^{+}, 2^{+}.2^{-}$ states. These results point out to the effect of restoring symmetries in the nuclear many body description of nuclear states. More details about these calculations are reported in [6] and work is in progress about ordinary muon capture on ^{76}Se [7]

Table 1. Energy states of ^{136}Cs. All values are given in units of keV.

J_n^π	Energy(Observed)	QRPA(g_{pp}=0.6)	PNQRPA(g_{pp}=0.3)
5^+	0.0	0.0	0.0
3_1^+	73.7	120	98
4_1^+	105.0	124	66
2_1^+	120 (?)	139	64
3_2^+	140.3	248	185
1^+	190 (?)	283	215
(4_2^+)	313.6	237	196
$(2^+, 3^+, 4^+)$	422.1-424.1	477	373
3^-	450(?)	475	453
8^-	517	447	453
1^+	588.8	515(*)	510(*)
2^-	700	872	678

Table 2. Muon capture rates to $J^\pi = 1^+$ states, as functions of the coupling g_{pp}. All values are given in units of $1/sec$.

g_{pp}	QRPA	QRPA (tbc)	PNQRPA	PNQRPA(tbc)
0.0	607632.62	324087.86	731535.86	38858.69
0.10	597070.69	318192.66	717884.87	380014.53
0.20	584425.69	311144.86	700311.56	369994.12
0.30	569606.94	302879.63	679222.69	357985.06
0.40	551239.93	292636.22	656492.87	344937.25
0.50	528064.00	279696.00	638736.69	334264.50
0.60	496966.78	262320.50	⋆	⋆
0.70	452329.06	237376.55	⋆	⋆
0.80	379049.47	196619.25	⋆	⋆

Table 3. Muon capture rates, in units of $1/sec$, for transitions leading to the 2^+ state of ^{136}Cs.

g_{pp}	QRPA	QRPA (tbc)	PNQRPA	PNQRPA(tbc)
0.0	245842.87	165671.28	224488.13	163432.79
0.10	233611.52	158038.17	228730.70	162149.28
0.20	220546.14	150484.31	223371.48	156082.14
0.30	206044.42	141829.17	211548.50	146918.94
0.40	189996.34	132298.98	195216.37	135615.33
0.50	172127.41	121670.02	176639.44	123427.15
0.60	152140.00	109782.73	159191.31	112338.31
0.70	129920.55	96591.57	151013.11	107404.52
0.80	105468.60	82105.79	⋆	⋆

Table 4. QRPA results, for the OMC rates with and without the inclusion of two body currents(tbc). The values are given in powers of ten and in units of $1/sec$. Only the dominant contributions are shown.´

J_n^π	Energy	QRPA	QRPA (tbc)
2_1^+	139	1.52(5)	1.09(5)
1_1^+	283	4.96(5)	2.62(5)
2_1^-	872	9.37(4)	5.11(4)

Table 5. The same (see the caption to Table 4) for the particle number projected QRPA approach

J_n^{π}	E	PNQRPA	PNQRPA (tbc)
2_1^+	64	2.11(5)	1.47(5)
1_1^+	215	6.79(5)	3.57(5)
2_2^+	218	1.05(5)	6.03(4)
2_1^-	678	6.08(5)	3.31(5)

4 Conclusions

In this work we have computed the rates of OMC on ^{136}Ba. The capture rates have been computed by using the proton-neutron quasiparticle random-phase approximation and its particle-number projected extension, including chiral two-body meson-exchange currents. The results of the energy spectra and the capture rates to states below 1 MeV of excitation in ^{136}Cs, calculated with both the QRPA and PNQRPA, were compared with the experimental low-energy spectrum. The inclusion of chiral two-body currents reduces the capture rates by some (30-40) % on average and the summed capture rates below 1 MeV of excitation in ^{136}Cs is a sizeable portion of the total capture rate. Studies of this type are relevant for large scale experimental efforts like MONUMENT [9].

Acknowledgements

This work has been done in collaboration with Jouni Suhonen and Jenni Kotila (University of Jyvaskyla, Finland).
Work done with partial financial support from the grants PICT 40492 (ANPCyT), the PIP 2081 (CONICET) of Argentina and the University of Jyvaskyla, Finland.

References

[1] J. Suhonen, O. Civitarese. Physics Reports 300 (1998) 123.
[2] J. D. Vergados, H. Ejiri and F. Simkovic. Int. J. Mod. Phys. E 25 (2016)1630007.
[3] M. Morita and A. Fujii. Phys. Rev.118 (1960) 606.
[4] L. Jokiniemi, T. Miyagi, S. R. Stroberg, J. D. Holt, J. Kotila, J. Suhonen. Phys. Rev. 107 (2023) 014327.
[5] J Suhonen and O Civitarese. Journal of Physics G 39 (2012)085105.
[6] O. Civitarese, A. Faessler, J. Suhonen and X. R. Wu. Physics Letters B 251 (1990)333.
[7] O. Civitarese, J. Kotila and J. Suhonen.. Physics Letters B (submitted for publication).
[8] O. Civitarese, J. Kotila and J. Suhonen (to be published).
[9] The MONUMENT Collaboration. Proceedings of the MEDEX'22 Workshop, Prague, Czech Republic, June 13-17, 2022.

Beta decays of heavy baryons

Hugo Antonio García Márquez[1] and *Roelof* Bijker[1,*]

[1]Instituto de Ciencias Nucleares, Universidad Nacions of the IBal Autónoma de México, A.P. 70-543, 04510 México D.F., México

Abstract. Beta decay processes of heavy baryons with a single heavy quark, charm or bottom, are analyzed in the quark model.

1 Introduction

The last decades have witnessed the discovery of a large number of hadrons with heavy quarks, among others by the LHCb, CMS, CDF, Belle and Babar collaborations. This has motivated a large activity in the study of the properties of heavy baryons and mesons. So far, most studies have concentrated on the mass spectrum and strong and electromagnetic decay widths (see *e.g.* [1–3] and references therein).

In this contribution, we present a study of the weak couplings of heavy baryons with a single heavy quark which can either be a charm or bottom quark in the context of the quark model. First, we review the structure of the wave functions of heavy baryons which is followed by a study of the axial couplings for semi-leptonic decay processes $B_Q \to B'_Q + e^- + \bar{\nu}_e$.

2 Wave functions

Baryons with a single charm or bottom quark belong to a $SU(3)$ flavor sextet **6** or anti-triplet $\bar{\mathbf{3}}$ (see Fig. 1) [4, 5]. The flavor wave function of singly-heavy baryons is written as qqQ, in which we distinguish between the light quarks $q = u, d, s$ and the heavy quark $Q = c, b$. The flavor sextet is symmetric under the interchange of the two light quarks and consists of a three isospin multiplets, a triplet Σ_Q, a doublet Ξ'_Q and a singlet Ω_Q. The spin-flavor wave functions with maximum charge state and spin $\frac{1}{2}$ are given by

$$\begin{aligned}
\Sigma_Q &= \frac{1}{\sqrt{6}} uuQ (2 \uparrow\uparrow\downarrow - \uparrow\downarrow\uparrow - \downarrow\uparrow\uparrow) , \\
\Xi'_Q &= \frac{1}{2\sqrt{3}} (us + su)Q (2 \uparrow\uparrow\downarrow - \uparrow\downarrow\uparrow - \downarrow\uparrow\uparrow) , \\
\Omega_Q &= \frac{1}{\sqrt{6}} ssQ (2 \uparrow\uparrow\downarrow - \uparrow\downarrow\uparrow - \downarrow\uparrow\uparrow) . \qquad (1)
\end{aligned}$$

The flavor anti-triplet is antisymmetric under the unterchange of the two light quarks and contains an isospin singlet Λ_Q and a doublet Ξ_Q. The corresponding spin-flavor wave functions

*e-mail: bijker@nucleares.unam.mx

6

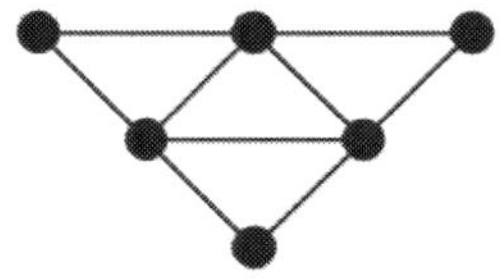

$\bar{\mathbf{3}}$

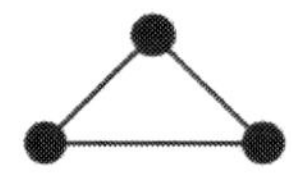

Figure 1. Single-heavy baryon flavor multiplets: sextet **6** and anti-triplet $\bar{\mathbf{3}}$. Here n denotes the non-strange light quarks $n = u, d$ and Q the heavy quarks $Q = c, b$.

with maximum charge state and spin $\frac{1}{2}$ are given by

$$\begin{aligned} \Lambda_Q &= \frac{1}{2}(ud - du)Q\,(\uparrow\downarrow\uparrow - \downarrow\uparrow\uparrow)\,, \\ \Xi_Q &= \frac{1}{2}(us - su)Q\,(\uparrow\downarrow\uparrow - \downarrow\uparrow\uparrow)\,. \end{aligned} \tag{2}$$

For the flavor $SU(3)$ we use the phase convention of Baird and Biedenharn [6].

3 Weak couplings

The baryon matrix elements of semi-leptonic decay processes $B_Q \to B'_Q + e^- + \bar{\nu}_e$ at low momentum transfer are given by the axial couplings [7]

$$g_A(B_Q \to B'_Q) = \left\langle B'_Q \middle| A_\mu \middle| B_Q \right\rangle , \tag{3}$$

where the axial current A_μ is given by

$$A_\mu = \bar{q}\,\gamma_\mu\gamma^5 \frac{1}{2}(\lambda_1 + i\lambda_2)\bar{q} = \bar{u}\,\gamma_\mu\gamma^5 d\,, \tag{4}$$

for $\Delta S = 0$, $\Delta Q = 1$ decays, and by

$$A_\mu = \bar{q}\,\gamma_\mu\gamma^5 \frac{1}{2}(\lambda_4 + i\lambda_5)\bar{q} = \bar{u}\,\gamma_\mu\gamma^5 s\,, \tag{5}$$

for $\Delta S = \Delta Q = 1$ decays. As an example, the axial coupling for the decay $\Sigma_c^+ \to \Sigma_c^{++} + e^- + \bar{\nu}_e$ can be derived as

$$\begin{aligned} g_A(\Sigma_c^+ \to \Sigma_c^{++}) &= \frac{1}{\sqrt{2}}\langle \Sigma_c^{++} | \,[A_\mu, I_-]\, | \Sigma_c^{++} \rangle = \frac{1}{\sqrt{2}}\langle \Sigma_c^{++} | \,\bar{u}\,\gamma_\mu\gamma^5 u - \bar{d}\,\gamma_\mu\gamma^5 d\, | \Sigma_c^{++} \rangle \\ &\to \frac{1}{\sqrt{2}}\langle \Sigma_c^{++} | \,2S_z(u) - 2S_z(d)\, | \Sigma_c^{++} \rangle = \frac{2\sqrt{2}}{3}\,. \end{aligned} \tag{6}$$

In the second line we have taken the nonrelativistic limit and used that the matrix elements of the spin operators are given by

$$\langle \Sigma_c^{++} | \,2S_z(q)\, | \Sigma_c^{++} \rangle = \begin{cases} \frac{4}{3} & q = u \\ 0 & q = d, s \\ -\frac{1}{3} & q = c \end{cases} \tag{7}$$

For the anti-triplet, we find in a similar way

$$g_A(\Xi_c^0 \to \Lambda_c^+) \quad \to \quad \langle \Lambda_c^+ | \, 2S_z(u) - 2S_z(s) \, | \Lambda_c^+ \rangle \; = \; 0 \, , \tag{8}$$

since in this case the spin of the Λ_c^+ baryon is carried entirely by the charm quark

$$\langle \Lambda_c^+ | \, 2S_z(q) \, | \Lambda_c^+ \rangle \; = \; \begin{cases} 0 & q = u, d, s \\ 1 & q = c \end{cases} \tag{9}$$

For the cross terms between the sextet and the anti-triplet one has

$$g_A(\Xi_c^{\prime 0} \to \Xi_c^+) \quad \to \quad \langle \Lambda_c^+ | \, 2S_z(u) - 2S_z(d) \, | \Xi_c^{\prime +} \rangle \; = \; -\frac{1}{\sqrt{3}} \, , \tag{10}$$

with

$$\langle \Lambda_c^+ | \, 2S_z(q) \, | \Xi_c^{\prime +} \rangle \; = \; \begin{cases} -\frac{1}{\sqrt{3}} & q = u \\ 0 & q = d, c \\ \frac{1}{\sqrt{3}} & q = s \end{cases} \tag{11}$$

In Table 1 we show the results for the weak decays among baryons of the sextet (top), the anti-triplet (middle) and between the sextet and the anti-triplet (bottom). In the quark model, the decays between members of the anti-triplet vanish identically. Since the single-charm and single-bottom baryons share the same dependence on the light quarks the axial couplings of Eqs. (3-5) are the same.

Table 1. Weak decays of singly heavy baryons

Decay				
$B_c \to B'_c$	$B_b \to B'_b$	g_A		CQM
$\Sigma_c^+ \to \Sigma_c^{++}$	$\Sigma_b^0 \to \Sigma_b^+$	$\frac{2\sqrt{2}}{3}$	=	0.943
$\Sigma_c^0 \to \Sigma_c^+$	$\Sigma_b^- \to \Sigma_b^0$	$\frac{2\sqrt{2}}{3}$	=	0.943
$\Xi_c^{\prime 0} \to \Xi_c^{\prime +}$	$\Xi_b^{\prime -} \to \Xi_b^{\prime 0}$	$\frac{2}{3}$	=	0.667
$\Xi_c^{\prime 0} \to \Sigma_c^+$	$\Xi_b^{\prime -} \to \Sigma_b^0$	$\frac{2}{3}$	=	0.667
$\Xi_c^{\prime +} \to \Sigma_c^{++}$	$\Xi_b^{\prime 0} \to \Sigma_b^+$	$\frac{2\sqrt{2}}{3}$	=	0.943
$\Omega_c^0 \to \Xi_c^{\prime +}$	$\Omega_b^- \to \Xi_b^{\prime 0}$	$\frac{2\sqrt{2}}{3}$	=	0.943
$\Xi_c^0 \to \Xi_c^+$	$\Xi_b^- \to \Xi_b^0$	0		
$\Xi_c^0 \to \Lambda_c^+$	$\Xi_b^- \to \Lambda_b^0$	0		
$\Sigma_c^0 \to \Lambda_c^+$	$\Sigma_b^- \to \Lambda_b^0$	$-\frac{\sqrt{2}}{\sqrt{3}}$	=	–0.816
$\Xi_c^{\prime 0} \to \Xi_c^+$	$\Xi_b^{\prime -} \to \Xi_b^0$	$-\frac{1}{\sqrt{3}}$	=	–0.577
$\Xi_c^{\prime 0} \to \Lambda_c^+$	$\Xi_b^{\prime -} \to \Lambda_b^0$	$-\frac{1}{\sqrt{3}}$	=	–0.577
$\Omega_c^0 \to \Xi_c^+$	$\Omega_b^- \to \Xi_b^0$	$-\frac{\sqrt{2}}{\sqrt{3}}$	=	–0.816

4 Summary and conclusions

Most studies of heavy baryons have concentrated on the mass spectrum and strong and electromagnetic decay widths. In this contribution, we presented a derivation of the axial couplings for the β decays of single-charm and single-bottom heavy baryons in the quark model.

The axial couplings are related by $SU(3)$ flavor symmetry. At the moment there is no experimental information to compare with. Future work will include the effects of the meson cloud on the axial couplings for heavy baryons.

Acknowledgments

This work was supported in part by PAPIIT-DGAPA grant IG101423 (Mexico).

References

[1] H.-X. Chen, W. Chen, X. Liu and S.-L. Zhu, Rep. Prog. Phys. **86**, 026201 (2023).
[2] P. Kloppenburg, "Particle properties", https://www.nikhef.nl/~kloppenb/particles.html (2024).
[3] S. Navas *et al.* (Particle Data Group Collaboration), Phys. Rev. D **110**, 030001 (2024).
[4] E. Ortiz-Pacheco and R. Bijker, Phys. Rev. D **108**, 054014 (2023).
[5] E. Ortiz-Pacheco and R. Bijker, arXiv:2410.09622 (2024).
[6] G.E. Baird and L.C. Biedenharn, J. Math. Phys. **4**, 1449 (1963).
[7] N. Cabibbo, E.C. Swallow and R. Winston, Annu. Rev. Nucl. Part. Sci. **53**, 39 (2003).

Updating the predictions for $X(3872)$ and $\psi(2S)$ production in heavy ion collisions to be measured by ALICE

Luciano M. Abreu[1,*], *Fernando* S. Navarra[2,**], and *Hildeson* P.L. Vieira[1,***]

[1]Instituto de Física, Universidade Federal da Bahia, Campus de Ondina, 40170-115, Bahia, Brazil
[2]Instituto de Física, Universidade de São Paulo, Rua do Matão, 1371, CEP 05508-090, São Paulo, Brazil

Abstract. We update previous predictions for the yield ratio $\mathcal{R} = N_{X(3872)}/N_{\psi(2S)}$ of particles produced at the LHC. We focus on the effects of the hadron gas phase on the abundance of these particles. We use a transport model to compute the multiplicities and how they evolve in time within the hadron gas. We find that the hadronic interactions yield a very weak suppression. Consequently, the ratio $\mathcal{R}$ for the tetraquark and meson molecule yields remains the same as the one at the hadronization time. In the present work we have improved the theoretical description of the $\psi(2S)$ interactions and used it to recalculate the ratio $\mathcal{R}$ in PbPb collisions at $\sqrt{s_{NN}} = 5.02$ TeV to be measured by the ALICE Collaboration in the Run 3. We find that the ratio obtained for the tetraquark configuration is 50 times smaller than for the molecular configuration. Within the uncertainties, this result is still consistent with our previous work.

1 Introduction

The existence of nonconventional, or exotic, hadronic states is by now well established. For a recent review, see [1]. However, we do not know yet which is their internal structure. In this work we consider two possible configurations: the meson molecule and the tetraquark.

The most well-known exotic state is the $X(3872)$. Its true nature remains controversial. After a few years of debate, we arrived at the conclusion that just studying the mass and decay width of the state will not be enough to unveil its structure [2]. It will be necessary, as pointed out by several groups, to study $X(3872)$ production in hadronic and nuclear collisions [3]. A few years ago the CMS Collaboration reported the first observation of $X(3872)$ in relativistic heavy-ion collisions [4]. The data [5] were given in terms of the ratio

$$\mathcal{R} = \frac{N_{X(3872)}}{N_{\psi(2S)}} \tag{1}$$

where N_i is the measured number of particles of species i. The best experimental set up to measure medium effects is provided by the ALICE Collaboration, which is expected to release data within one year from now.

* e-mail: luciano.abreu@ufba.br
** e-mail: navarra@if.usp.br
*** e-mail: hildeson.paulo@ufba.br

In the late stage of a high energy nucleus-nucleus collison a hadron gas is formed in which the particles can undergo multiple scatterings. They can be destroyed in absorption processes but can also be formed in the inverse processes. The final multiplicity will depend ultimately on the spatial configuration of the states.

In [6], using a transport model, we studied particle production in PbPb collisions at $\sqrt{s_{NN}} = 5.02$ TeV. We made the following predictions for the ratio $\mathcal{R}$:

$$\mathcal{R} \simeq 5 \quad \text{for molecules} \qquad \mathcal{R} \simeq 0.1 \quad \text{for tetraquarks} \tag{2}$$

After the publication of our paper, a new article from the ALICE Collaboration on the ratio $\mathcal{R}_\psi = N_{\psi(2S)}/N_{J/\psi}$ appeared [7]. These results were extremely relevant for the study of $\psi(2S)$ interactions with light mesons and could yield new and important constraints on our model of the $\psi(2S)$ interactions. We then decided to address the data of [7]. This led us to review some assumptions concerning the coupling constants involving the $\psi(2S)$ and change them. With new coupling constants we were able to reproduce the data of [7]. Our results are contained in the article [8]. The impact of changing the $\psi(2S)$ coupling constants can be seen in Fig. 1. In the figure, on the left panel we see the results of our previous model (used in [6]) and on the right panel the results obtained with the updated model. The natural next step is to go back to the calculations presented in [6], change the $\psi(2S)$ coupling constants, redo the calculations and update the predictions for $\mathcal{R}$. This will be done here. In what follows we will briefly describe the steps of the transport model and the introduced changes.

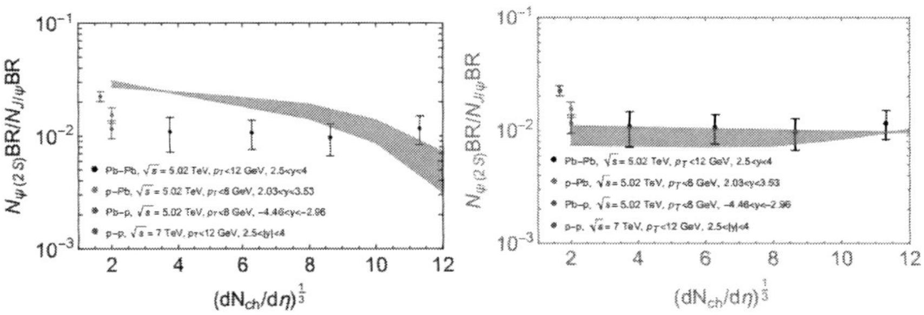

(a) (b)

Figure 1. Ratio $\mathcal{R}_\psi$ as a function of the system size. Data are from [7] and bands are theoretical calculations. a) Theoretical results obtained with the $\psi(2S)$ interaction model of [6]. b) Theoretical results obtained with the $\psi(2S)$ interaction model of [8].

2 Interactions in the hadronic medium

In [9] we revisited the formalism developed in [10] to study the interactions of multiquark states with light mesons in a hadronic medium. In the case of the $J/\psi - \pi$ we obtained results which are consistent with those found previously with QCD sum rules [11]. We assumed that the $\psi(2S)$ interactions are described by the same Lagrangians used for the J/ψ with the form factors calculated in [12] but with different coupling constants. It is well known that at high energies the cross sections tend to approach the geometric limit. Following this high energy approximation [6] we assumed that in the interactions with other mesons, the charmonium cross sections, both $\sigma_{J/\psi}$ and $\sigma_{\psi(2S)}$, are proportional to the square of the radius of the charmonium state. Since $r_{\psi(2S)} \simeq 2\, r_{J/\psi}$, we used coupling constants varying in the range $[g, 2g]$. Using these cross sections as input in our transport model, we could calculate the ratio $N_{\psi(2S)}/N_{J/\psi}$ as a function of the system size $\mathcal{N} = [dN_{ch}/d\eta(|\eta| < 0.5)]^{1/3}$. The results are shown in Fig. 1a. They are compared to the experimental data of Ref. [7]. Clearly, there is no agreement between theory and data. Since the other ingredients of our transport model had been successfully tested in other calculations, our most natural reaction would be to change the newly introduced $\psi(2S)$ coupling constants.

After the publication of [7] we revisited the literature on $\psi(2S)$ interactions. We concluded that the couplings between $\psi(2S)$ and other mesons are essentialy unknown and the existing estimates do not converge. On one extreme we have the results obtained in [13], where the authors conclude that in decays into a vector (V) and a pseudoscalar (P) meson the relation bewteen the couplings is $g_{\psi' VP} \simeq 0.13\, g_{J/\psi VP}$. In [14], using heavy quark effective theory, the authors found that the coupling constant of the $\psi(nS)$ state with the $D\bar{D}$ mesons decreases with n. In [15] the authors found that $g_{\psi' DD} = 0.9\; g_{J/\psi DD}$ and also $g_{\psi' D^* D^*} = 0.9\; g_{J/\psi D^* D^*}$. In [16] in the study of the $e^+e^- \to D\bar{D}$ cross section lineshape, it was found that $g_{\psi' DD} \simeq g_{J/\psi DD}$. Finally, in [17] it was found that $g_{\psi' DD} = 1.67\; g_{J/\psi DD}$. Since we still do not have a definite calculation or a measurement for the $\psi(2S)$ couplings, we take their values in the range

$$0.4 \leq g_{\psi' M_1 M_2}/g_{J/\psi M_1 M_2} \leq 0.5 \tag{3}$$

This ratio lies in the range of the phenomenological predictions ($0.13 \leq g_{\psi' M_1 M_2}/g_{J/\psi M_1 M_2} \leq 1.67$) and leads to a good agreement with the data. Repeating the steps of the previous calculation, we used the cross sections obtained with (3) as input in the transport model and calculated the ratio $N_{\psi(2S)}/N_{J/\psi}$. The results are shown in Fig. 1b. As it can be seen, we find a very good agreemente between our model and the data. It is important to emphasize that the initial conditions were the same in both calculations.

3 Time evolution

Using the Lagrangians listed in [6] we could calculate the appropriate cross sections, solve the rate equation and compute the time evolution of the hadron multiplicities. The rate equation reads:

$$\frac{dN_h(\tau)}{d\tau} = \sum_{a,b,c} \left[\langle\sigma_{ab\to ch} v_{ab}\rangle n_a(\tau) N_b(\tau) - \langle\sigma_{ch\to ab} v_{ch}\rangle n_c(\tau) N_h(\tau)\right], \tag{4}$$

where $N_h(\tau)$ is the multiplicity of the state h ($h = \psi(2S), X(3872)$); $n_i(\tau)$ and $N_i(\tau)$ represent the density and the number of the meson of type i at a given time τ, respectively. The thermal cross section $\langle\sigma_{ab\to cd} v_{ab}\rangle$ of the process $ab \to cd$ is given by:

$$\langle\sigma_{ab\to cd} v_{ab}\rangle = \frac{\int d^3\mathbf{p}_a d^3\mathbf{p}_b f_a(\mathbf{p}_a) f_b(\mathbf{p}_b) \sigma_{ab\to cd}\; v_{ab}}{\int d^3\mathbf{p}_a d^3\mathbf{p}_b f_a(\mathbf{p}_a) f_b(\mathbf{p}_b)} \tag{5}$$

where v_{ab} denotes the relative velocity of the two initial interacting particles and the function $f_i(\mathbf{p}_i)$ is the Bose-Einstein distribution. The particles of the hadron gas follow the distribution

$$n_i(\tau) = \frac{1}{2\pi^2} \gamma_i g_i m_i^2 T(\tau) K_2\left(\frac{m_i}{T(\tau)}\right), \tag{6}$$

where γ_i, g_i and m_i is the fugacity, degeneracy factor and mass of the particle of type i, respectively. The multiplicity $N_i(\tau)$ is then given by:

$$N_i(\tau) = n_i(\tau) \times V(\tau) \tag{7}$$

According to the model presented in [18] the time evolution of the system is given by:

$$
\begin{aligned}
V(\tau) &= \pi\left[R_C + v_C(\tau - \tau_C) + \frac{a_C}{2}(\tau - \tau_C)^2\right]^2 \tau c, \\
T(\tau) &= T_C - (T_H - T_F)\left(\frac{\tau - \tau_H}{\tau_F - \tau_H}\right)^{\frac{4}{5}}, \qquad (8)
\end{aligned}
$$

where R_C, v_C, a_C and T_C are the transverse size, transverse velocity, transverse acceleration and temperature at the time τ_C, respectively; $T_H(T_F)$ is the temperature at the hadronization (kinetic freeze-out) time $\tau_H(\tau_F)$. The parameters in Eq. (8) are fixed for a hadronic medium formed in central PbPb collisions at $\sqrt{s_{NN}} = 5.02$ TeV [18].

4 Multiplicities

Hadronization can be described by the coalescence model (CM) [18] and by the statistical hadronization model (SHM) [19]. From the coalescence model one obtains the following multiplicities for $\psi(2S)$ and $X(3872)$ at the end of the mixed phase [6, 18]:

$$
N_{\psi(2S)}(\tau_H) \approx 1.8 \times 10^{-4}, \qquad N_X^{(4q)}(\tau_H) \approx 1.8 \times 10^{-4}, \qquad N_X^{(Mol)}(\tau_H) \approx 1.1 \times 10^{-2}. \quad (9)
$$

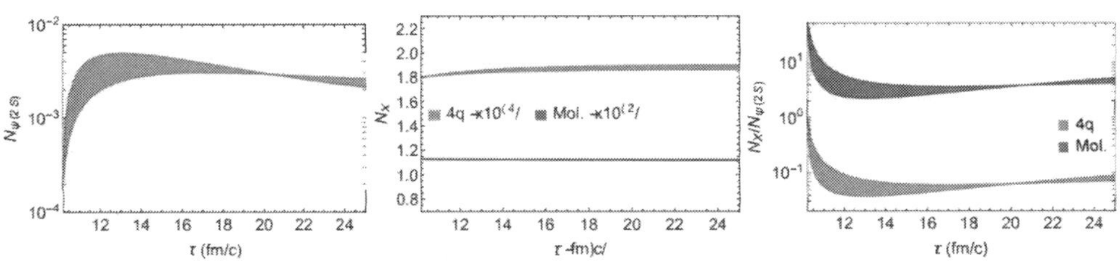

(a) (b) (c)

Figure 2. Time evolution of different quantities in central PbPb collisions at $\sqrt{s_{NN}} = 5.02$ TeV. a) Multiplicity of $\psi(2S)$. b) Multiplicity of $X(3872)$ for different configurations. c) Ratio of the $X(3872)$ and $\psi(2S)$ for different configurations.

With these initial conditions, Eq. (4) can be solved. We show the results in Fig. 2. The bands illustrate the effect of the uncertainty in the $\psi(2S)$ couplings, given in (3). In Fig. 2a we show the time evolution of the $\psi(2S)$ yield. In Fig. 2b we see that the abundance of tetraquarks and molecules, is nearly conserved during the hadron gas phase. In Fig. 2c one can see how the ratio $\mathcal{R}$ changes rapidly in the beginning and after that tends to reach an asymptotic value.

5 System size dependence

The size of the system is proportional to the charged-particle pseudorapidity density at mid-rapidity, $\mathcal{N} = [dN_{ch}/d\eta](|\eta| < 0.5)$, which can be connected to the freeze-out temperature by means of the empirical relation [20, 21]:

$$
T_F = T_{F0}\, e^{-b\mathcal{N}}, \qquad (10)
$$

where T_{F0} and b are constants. Following [20, 21] we assume that

$$T = T_h\,(\tau_h/\tau)^{1/3}$$

and hence the freeze-out time τ_F becomes a function of $\mathcal{N}$:

$$\tau_F = \tau_H \left(\frac{T_H}{T_{F0}}\right)^3 e^{3b\mathcal{N}}. \tag{11}$$

A larger source implies a bigger $\mathcal{N}$ and thus a a bigger τ_F. Using Eq. (11) in (4) introduces a dependence of the multiplicity N_h on the size of the source. In Ref. [22], it was shown that the volume of the system (V), the charm quark number (N_c) and light quark number (N_q) can be written in terms of $\mathcal{N}$ through similar empirical formulas. They are [22]:

$$V = 2.82\,\mathcal{N}^3 \qquad N_c = 7.9\times10^{-5}\,\mathcal{N}^{4.8} \qquad N_u = N_d = 0.37\,\mathcal{N}^3. \tag{12}$$

Using the above expressions in the coalescence model formula to compute $N_{X(3872)}$ and $N_{\psi(2S)}$, it is possible to obtain their ratio, $\mathcal{R}$. The results are shown in Fig. 3. It is important to mention that the points on the left part of the plots shown in Fig. 3 (where $2 < \mathcal{N} < 4$) are at the limit of the validity of our approach, since in this region the system becomes very small and the existence of a fluid is doubtful.

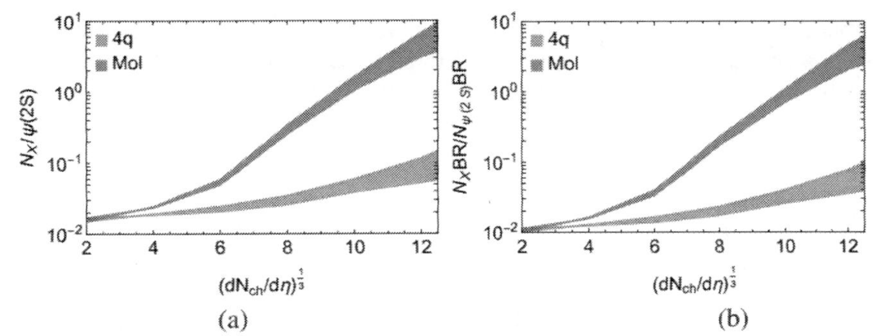

Figure 3. a) Ratio $\mathcal{R}$ as a function of the system size $\mathcal{N}$. b) Ratio $\mathcal{R}$ corrected by the decay branching ratios of $X(3872)$ and $\psi(2S)$ as a function of the system size $\mathcal{N}$.

6 Discussion

The results shown in Fig. 3a, suggest that in central PbPb collisions at $\sqrt{s_{NN}} = 5.02$ TeV, where $\mathcal{N} = 12.4$, the ratio $\mathcal{R}$ for minimum bias events is:

$$\frac{N_X}{N_{\psi(2S)}} \simeq 6.1 \pm 2.6 \quad \text{molecules} \qquad \frac{N_X}{N_{\psi(2S)}} \simeq 0.1 \pm 0.04 \quad \text{tetraquarks} \tag{13}$$

These numbers should be compared with (2). As it can be seen, the two sets of numbers are compatible with each other, with a slight increase in the number of molecules. The difference is that now our model is also compatible with the $\psi(2S)$ production data, as shown in Fig.

1b. Moreover, now we have an estimate of the theoretical error bars. The predictions for $\mathcal{R}$ will be tested at the run 3 of the ALICE Collaboration, where it will be possible to measure $X(3872)$ and $\psi(2S)$ at low transverse momentum, i.e. $p_T < 12$ GeV. The data will most likely be given in terms of the mutiplicities multiplied by the decay branching ratios, as done in [4, 5, 7]. Therefore it is useful to plot the branching ratio corrected yield ratio, which is shown in Fig. 3b. From the figure we can read the corrected ratios, which are:

$$\frac{N_X\,BR}{N_{\psi(2S)}\,BR} \simeq 4.1 \pm 1.7 \quad \text{molecules} \qquad \frac{N_X\,BR}{N_{\psi(2S)}\,BR} \simeq 0.065 \pm 0.025 \quad \text{tetraquarks} \tag{14}$$

In view of the numbers above it seems that it is indeed possible to discriminate between molecules and tetraquarks. Our results also indicate that it is easier to make this discrimination in larger systems and in more central collisions.

Before closing this section, a comment on the CMS data [4] is in order. The data were collected in PbPb collisions at a center-of-mass energy $\sqrt{s} = 5.02$ TeV per nucleon pair, using the decay chain $X(3872) \to J/\psi\pi^+\pi^- \to \mu^+\mu^-\pi^+\pi^-$. The rapidity and transverse momentum intervals considered were $|y| < 1.6$ and $15 < p_T < 50$ GeV. The prompt $X(3872)$ to $\psi(2S)$ yield ratio was found to be:

$$\mathcal{R} \;=\; \frac{N_{X(3872)}}{N_{\psi(2S)}} = 1.08 \pm 0.49(\text{stat}) \pm 0.52(\text{syst}). \tag{15}$$

Comparing this number with (14), we see that they do not agree with each other. This is not surprising and it is a consequence of the transverse momentum cut used by the CMS Collaboration. Indeed, particles produced with these values of p_T are not likely to come from the hadronization of the quark gluon plasma. Probably they were produced at early stages of the collisions and did not experience all the interactions discussed here. This is why a direct comparison between our results and CMS data is not meaningful. We will have to wait for the ALICE data, which, measuring low p_T particles will bring information about the quark gluon plasma and about the hadron gas.

References

[1] Z. G. Wang, arXiv:2502.11351; H. X. Chen, W. Chen, X. Liu, Y. R. Liu and S. L. Zhu, Rept. Prog. Phys. **86**, 026201 (2023).
[2] S. Narison, F. S. Navarra and M. Nielsen, Phys. Rev. D **83**, 016004 (2011).
[3] S. Cho *et al.* [ExHIC], Phys. Rev. Lett. **106**, 212001 (2011).
[4] A. M. Sirunyan *et al.* [CMS], Phys. Rev. Lett. **128**, 032001 (2022).
[5] R. Aaij *et al.* [LHCb]; Phys. Rev. Lett. **132**, 242301 (2024).
[6] L. M. Abreu, F. S. Navarra and H. P. L. Vieira, Phys. Rev. D **110**, 014011 (2024).
[7] S. Acharya *et al.* [ALICE], Phys. Rev. Lett. **132**, 042301 (2024).
[8] L. M. Abreu, F. S. Navarra and H. P. L. Vieira, arXiv:2409.12755.
[9] A. M. Torres, K. P. Khemchandani, F. S. Navarra, M. Nielsen, L. M. Abreu, Phys. Rev. D **90**, 114023 (2014); L. M. Abreu, K. P. Khemchandani, A. Martinez Torres, F. S. Navarra, and M. Nielsen, Phys. Lett. B **761**, 303 (2016).
[10] S. Cho and S. H. Lee, Phys. Rev. C **88**, 054901 (2013).
[11] F. S. Navarra, M. Nielsen, R. S. Marques de Carvalho and G. Krein, Phys. Lett. B **529**, 87 (2002); F. O. Duraes, S. H. Lee, F. S. Navarra and M. Nielsen, Phys. Lett. B **564**, 97 (2003).
[12] M. E. Bracco, M. Chiapparini, F. S. Navarra and M. Nielsen, Prog. Part. Nucl. Phys **67**, 1019 (2012).

[13] G. Li, Q. Zhao and C. H. Chang, J. Phys. G **35**, 055002 (2008).
[14] S. Liu, Z. Cai, Y. Zheng and G. Li, Eur. Phys. J. C **83**, 820 (2023).
[15] G. Li and Q. Zhao, Phys. Rev. D **84**, 074005 (2011).
[16] Y. J. Zhang and Q. Zhao, Phys. Rev. D **81**, 034011 (2010).
[17] Y. Dong, A. Faessler, T. Gutsche and V. E. Lyubovitskij, J. Phys. G **38**, 015001 (2011).
[18] S. Cho *et al.* [EXHIC], Prog, Part. Nucl. Phys. **95**, 279 (2017).
[19] A. Andronic, P. Braun-Munzinger, M. K. Köhler, K. Redlich and J. Stachel, Phys. Lett. B **797**, 134836 (2019).
[20] B. Abelev et al. [ALICE Collaboration], Phys.Rev. C **88**, 044910 (2013).
[21] C. Le Roux, F. S. Navarra and L. M. Abreu, Phys. Lett.B **817**, 136284 (2021).
[22] L. M. Abreu, H. P. L. Vieira and F. S. Navarra, Phys. Rev. D **105**, 116029 (2022).

EPJ Web of Conferences **333**, 02004 (2025) https://doi.org/10.1051/epjconf/202533302004
XLVI Symposium on Nuclear Physics 2025

The use of many-body methods for the calculation of meson-like spectra.

O.A. Rico-Trejo[1*], *T* Yépez-Martínez[2**], *P. O.* Hess[1,3***], and *O.* Civitarese.[4****]

[1]Instituto de Ciencias Nucleares, Ciudad Universitaria, Mexico-City, México.
[2]Instituto de Educación Media Superior de la Ciudad de Mexico, Mexico-City, México.
[3]Frankfurt Institute for Advanced Studies, J.W. Von Goethe University, Hessen, Germany.
[4]Departamento de Física, Universidad Nacional de la Plata, La Plata, Argentina.

Abstract. The study of properties for the low-energy regime of hadronic spectra is a challenging task; e.g. meson-like states require the introduction of effective degrees of freedom, a condition resulting from confinement. This work presents the results of adopting a non-perturbative scheme where quarks are treated as quasiparticles. These quasiparticles interact and by employing many-body techniques, we can describe a spectrum of meson-like states as a collective superposition of quasiquark-pairs followed by calculating their widths. The result of the calculations shows that this scheme is a suitable one to describe meson states up to energies of the order of a couple of GeV.

1 Introduction

The identification of effective degrees of freedom in a physical theory, such as the many-body problem in the quantum regime, served for many years as the main guide by describing physical systems in the QCD low-energy regime, which is a highly non-perturbative theory. In previous works, The SO(4) symmetry model was proposed in [1], which allowed a description of meson-like particles' effective degrees of freedom in addition to calculating their widths.

In recent years, the Cornell-like potential, composed by a Coulomb interaction ($\frac{1}{r}$) plus a linear potential (r), has been considered a good choice to describe the quark confinement [2, 3]. In this work, motivated by previous efforts [4, 5], we aimed to provide a global description of the meson-like spectra properties in the nuclear many-body framework from a single-fit of parameters. We constructed an effective Hamiltonian starting from the QCD Hamiltonian in the Coulomb gauge [6] considering a pure quark-confined interaction. The harmonic oscillator basis was used to describe the particle states, afterwards a transformation of particle basis to a quasiparticle representation and the BCS method were used to obtain the mass. Finally, the description of properties of the meson-like states by the quasiparticles interaction is carried out with the Random Phase Approximation (RPA) method [7], and if relevant, the calculations of their widths.

*e-mail: octavio.augusto@ciencias.unam.mx
**e-mail: tochtlicuauhtli.yepez@iems.edu.mx
***e-mail: hess@nucleares.unam.mx
****e-mail: osvaldo.civitarese@gmail.com

2 Effective Hamiltonian and canonical transformation.

The proposed effective Hamiltonian for the QCD low-energy regimen restricted to the quark sector with a confined interaction $V(|\mathbf{x}-\mathbf{y}|) = -\frac{\alpha_c}{V|\mathbf{x}-\mathbf{y}|} + \beta_L V|\mathbf{x}-\mathbf{y}|$ is:

$$\begin{aligned}\mathcal{H}^{Eff}_{QCD} &= \int \{\bar{\psi}(-i\alpha\cdot\nabla + \beta m)\psi\} d\mathbf{x} - \frac{1}{2}\int \rho_c(\mathbf{x})V(|\mathbf{x}-\mathbf{y}|)\rho^c(\mathbf{y})d\mathbf{x}d\mathbf{y}. \\ &= K + \mathcal{H}_{Coul}, \end{aligned} \tag{1}$$

where the fermionic field $\psi^\dagger(\mathbf{x})$ is expanded in terms of the harmonic oscillator basis, α_c is the associated parameter to Coulomb interaction and β_L is the parameter of string tension.

As a result of writing the fields, the effective kinetic term of the Hamiltonian $\mathcal{H}^{Eff}_{QCD}$ in the Eq. (1) is no longer diagonal, so we must diagonalize it. We will refer to this process as *prediagonalization*. Also, using this basis creates a new way of writing the terms of the Hamiltonian for effective quark $(b^\dagger, b)$ and antiquark $(d^\dagger, d)$ degrees of freedom. The kinetic term is rewritten as

$$\mathbf{K} = \sum_{\pi k\gamma} \epsilon_{\lambda\pi k\gamma} \sum_{\mu} \left(\mathbf{b}^\dagger_{\pi k\gamma\mu}\mathbf{b}_{\pi k\gamma\mu} - \mathbf{d}^{\pi k\gamma\mu}\mathbf{d}^\dagger_{\pi k\gamma\mu}\right), \tag{2}$$

where the short-hand notation $\gamma = (J, c, (Y, T))$ and $\mu = (m, m_c, T_z)$ is used and $\epsilon_{\lambda\pi k\gamma}$ represents the eigenvalue of the effective quark energies [8].

The pairlike terms in the Hamiltonian acquire a form

$$\begin{aligned}\mathcal{F}_{\lambda_1\mathbf{q}_1,\lambda_2\mathbf{q}_2;\gamma_{f_0},\mu_{f_0}} &= \frac{1}{\sqrt{2}}\left\{\delta_{\lambda_{1,\frac{1}{2}}}\delta_{\lambda_{2,\frac{1}{2}}}\left[\mathbf{b}^\dagger_{\mathbf{q}_1}\otimes\mathbf{b}_{\bar{\mathbf{q}}_2}\right]^{\gamma_{f_0}}_{\mu_{f_0}} - \delta_{\lambda_{1,-\frac{1}{2}}}\delta_{\lambda_{2,-\frac{1}{2}}}\left[\mathbf{d}_{\mathbf{q}_1}\otimes\mathbf{d}^\dagger_{\bar{\mathbf{q}}_2}\right]^{\gamma_{f_0}}_{\mu_{f_0}}\right\} \\ \mathcal{G}_{\lambda_1\mathbf{q}_1,\lambda_2\mathbf{q}_2;\gamma_{f_0},\mu_{f_0}} &= \frac{1}{\sqrt{2}}\left\{\delta_{\lambda_{1,-\frac{1}{2}}}\delta_{\lambda_{2,\frac{1}{2}}}\left[\mathbf{d}_{\mathbf{q}_1}\otimes\mathbf{b}_{\bar{\mathbf{q}}_2}\right]^{\gamma_{f_0}}_{\mu_{f_0}} - \delta_{\lambda_{1,\frac{1}{2}}}\delta_{\lambda_{2,-\frac{1}{2}}}\left[\mathbf{b}^\dagger_{\mathbf{q}_1}\otimes\mathbf{d}^\dagger_{\bar{\mathbf{q}}_2}\right]^{\gamma_{f_0}}_{\mu_{f_0}}\right\}. \end{aligned} \tag{3}$$

We also have used a short-hand notation $1 = \lambda_1\pi_1k_1J_1Y_1T_1$ (same for index 2) for the quantum numbers of the intermediate coupling. With this, the Coulomb interaction acquires the following structure

$$\begin{aligned}\mathcal{H}_{Coul} = -\frac{1}{2}\sum_{L}\sum_{\lambda_i\mathbf{q}_i} \mathbf{V}^L_{(\lambda_i\mathbf{q}_i)}\Big(&\left[\mathcal{F}_{\lambda_1\mathbf{q}_1,\lambda_2\mathbf{q}_2;\gamma_{f_0}}\mathcal{F}_{\lambda_3\mathbf{q}_3,\lambda_4\mathbf{q}_4;\bar{\gamma}_{f_0}}\right]^{\gamma_0}_{\mu_0} + \left[\mathcal{F}_{\lambda_1\mathbf{q}_1,\lambda_2\mathbf{q}_2;\gamma_{f_0}}\mathcal{G}_{\lambda_3\mathbf{q}_3,\lambda_4\mathbf{q}_4;\bar{\gamma}_{f_0}}\right]^{\gamma_0}_{\mu_0} \\ &+ \left[\mathcal{G}_{\lambda_1\mathbf{q}_1,\lambda_2\mathbf{q}_2;\gamma_{f_0}}\mathcal{F}_{\lambda_3\mathbf{q}_3,\lambda_4\mathbf{q}_4;\bar{\gamma}_{f_0}}\right]^{\gamma_0}_{\mu_0} + \left[\mathcal{G}_{\lambda_1\mathbf{q}_1,\lambda_2\mathbf{q}_2;\gamma_{f_0}}\mathcal{G}_{\lambda_3\mathbf{q}_3,\lambda_4\mathbf{q}_4;\bar{\gamma}_{f_0}}\right]^{\gamma_0}_{\mu_0}\Big), \end{aligned} \tag{4}$$

The following section describes the transformations at the quasiparticle basis in terms of the effective quark operators and the BCS equations and their solutions and, results for the energy and gap for each quasiparticle type.

3 Bogoliubov transformations basis and BCS equations

The pair of terms of the Eq. (3) ought to be written in the basis of quasiparticle operators by a Bogoliubov transformation, which leads for the creation operators

$$\begin{aligned} B^\dagger_{k_i\pi_i,\gamma_i\mu_i} &= u_{k_i\pi_i,\gamma_i}\mathbf{b}^\dagger_{k_i\pi_i,\gamma_i\mu_i} - v_{k_i\pi_i,\gamma_i}\mathbf{d}_{k_i\pi_i,\gamma_i\mu_i} \\ D^{\dagger k_i\pi_i,\gamma_i\mu_i} &= u_{k_i\pi_i,\gamma_i}\mathbf{d}^{\dagger k_i\pi_i,\gamma_i\mu_i} + v_{k_i\pi_i,\gamma_i}\mathbf{b}^{k_i\pi_i,\gamma_i\mu_i}, \end{aligned} \tag{5}$$

and the annihilation operators

$$
\begin{aligned}
B^{k_i\pi_i,\gamma_i\mu_i} &= u^*_{k_i\pi_i,\gamma_i}\mathbf{b}^{k_i\pi_i,\gamma_i\mu_i} - v^*_{k_i\pi_i,\gamma_i}\mathbf{d}^{\dagger k_i\pi_i,\gamma_i\mu_i} \\
D_{k_i\pi_i,\gamma_i\mu_i} &= u^*_{k_i\pi_i,\gamma_i}\mathbf{d}_{k_i\pi_i,\gamma_i\mu_i} + v^*_{k_i\pi_i,\gamma_i}\mathbf{b}^{\dagger}_{k_i\pi_i,\gamma_i\mu_i}.
\end{aligned} \tag{6}
$$

In both cases, the elements $u_{k_i\pi_i,\gamma_i}$ and $v_{k_i\pi_i,\gamma_i}$ are the factors of the Bogoliubov transformations between quarks and quasiquarks operators. It's worth mentioning the bare vacuum of the quark will be replaced by a correlated vacuum of quasiparticles which fulfills the condition $B|BCS\rangle = 0$ and $D|BCS\rangle = 0$

The next step consists of transforming both terms of the Eq. (2) and Eq. (4) of the Hamiltonian to the quasiquark basis in order to write and solve the BCS equations for occupation factors and gaps, for each of the channels entering in these expressions. That step is quite long and we will omit it here, the details are disclosed in the main work at [9]. The BCS procedure requires that the expectation value of the terms with two creation or two annihilation quasiquark operators should vanish when acting on the correlated vacuum. Together with the condition that the term with one quasiquark creation and annihilation operators should be diagonal when acting on the correlated vacuum. These requirements lead to the determination of the values of the gap and occupation factor for each quark flavor. The resulting set of equations, using the short-hand notation $\kappa = k\pi\gamma$, is

$$
\begin{aligned}
\Sigma_{\kappa_1} Z_{\kappa_1} + \Delta_{\kappa_1} W_{\kappa_1} &= E_{\kappa_1} \\
-\Delta_{\kappa_1} Z_{\kappa_1} + \Sigma_{\kappa_1} W_{\kappa_1} &= 0
\end{aligned} \tag{7}
$$

where the E_{κ_1} are the quasiquark energies and

$$
\begin{aligned}
\Sigma_{\kappa_1} &= \epsilon_{\kappa_1} + \bar{V}^{\Sigma}_{\kappa_1\kappa_2\kappa_3\kappa_4}(u^2_{\kappa_2} - v^2_{\kappa_2}) \\
\Delta_{\kappa_1} &= \bar{V}^{\Delta}_{\kappa_1\kappa_2\kappa_3\kappa_4}(u_{\kappa_2}v_{\kappa_2}) \\
E_{\kappa_1} &= \sqrt{\Sigma^2_{\kappa_1} + \Delta^2_{\kappa_1}} \\
Z_{\kappa_1} &= \left\{u^2_{\kappa_1} - v^2_{\kappa_1}\right\} \\
W_{\kappa_1} &= \left\{2u_{\kappa_1}v_{\kappa_1}\right\}.
\end{aligned} \tag{8}
$$

The last step which will lead us to the meson-like spectrum construction is, the application of the many-body method Random Phase Approximation (RPA) and, in case the state's identification might not be clear, use of the Bohr-Mottelson method to calculate the decay with of some particular states.

4 The random phase approximation procedure

Once we have transformed the Hamiltonian with the terms of the Eq. (2) and Eq. (4) to the quasiparticle basis, there remain some terms with the creation and annihilation of one and two pairs of coupled quasiquarks. This Hamiltonian, which will be called $\mathcal{H}^{QCD}_{RPA}$ contains a term of creation and annihilation of quasiparticle pairs $\mathcal{H}_{22}$ and a term of creation or annihilation of two quasiparticle pairs $\mathcal{H}_{40}, \mathcal{H}_{04}$, respectively, besides a number quasiparticle operator $\mathcal{H}_{11}$. The RPA method consists of the diagonalization of the Hamiltonian

$$\mathcal{H}^{QCD}_{RPA} = \mathcal{H}_{11} + \mathcal{H}_{22} + \mathcal{H}_{40} + \mathcal{H}_{04}, \tag{9}$$

in a new basis of phonons conformed by a linear combination of quasiquark pair-creation and quasiquark pair-annihilation operators. These phonons are defined as

$$\hat{\Gamma}^{\dagger}_{n;\Gamma\mu} = \sum_{\mathbf{a},\mathbf{b}} X^{n}_{\mathbf{ab};\Gamma}[B^{\dagger}_{\mathbf{a}}D^{\dagger}_{\mathbf{\bar{b}}}]^{\Gamma}_{\mu} - Y^{n}_{\mathbf{ab};\Gamma}(-1)^{\phi_{\Gamma\mu}}[D^{\mathbf{\bar{b}}}B^{\mathbf{a}}]^{\bar{\Gamma}}_{\bar{\mu}}, \tag{10}$$

where the coefficients $X^{n}_{\mathbf{ab};\Gamma}$ and $Y^{n}_{\mathbf{ab};\Gamma}$ are the amplitudes of the creation-annihilation pairs, respectively. The dynamic equation can be written as the double commutator

$$\langle RPA|\left[\hat{\Gamma}^{n';\Gamma\mu},\left[H^{QCD}_{RPA},\hat{\Gamma}^{\dagger}_{n;\Gamma\mu}\right]\right]|RPA\rangle = E^{RPA}_{n;\Gamma}\delta_{n,n'}. \tag{11}$$

The quantities $E^{RPA}_{n;\Gamma}$ are the *nth* one-phonon eigenvalues to be associated with the mesonic states. In the same way we did it in the BCS, we have to define a new vacuum state that fulfills the condition of $\hat{\Gamma}^{n';\Gamma\mu}|RPA\rangle = 0$.

The general solution of the RPA method requires the solving of two sets of equations from the amplitudes $X^{n}_{\mathbf{ab};\Gamma}$ and $Y^{n}_{\mathbf{ab};\Gamma}$. These can be written in matrix form as

$$\begin{pmatrix} A & B \\ B^{*} & A^{*}\end{pmatrix}\begin{pmatrix} X^{n} \\ Y^{n}\end{pmatrix} = E^{RPA}_{n}\begin{pmatrix} 1 & 0 \\ 0 & -1\end{pmatrix}\begin{pmatrix} X^{n} \\ Y^{n}\end{pmatrix}, \tag{12}$$

the matrices A and B are the so-called forward and backward matrices and can be written, in the quasiboson approximation scheme, as

$$\begin{aligned} A_{a'b';\Gamma'\mu';\,a,b;\Gamma\mu} &= \langle\tilde{0}|\left[[D^{\mathbf{\bar{b}}'}B^{\mathbf{a}'}]^{\Gamma'}_{\mu'},\left[H^{QCD}_{RPA},[B^{\dagger}_{\mathbf{a}}D^{\dagger}_{\mathbf{\bar{b}}}]^{\Gamma}_{\mu}\right]\right]|\tilde{0}\rangle\,, \\ B_{a'b';\Gamma'\mu';\,a,b;\Gamma\mu} &= -\langle\tilde{0}|\left[[D^{\mathbf{\bar{b}}'}B^{\mathbf{a}'}]^{\Gamma'}_{\mu'},\left[H^{QCD}_{RPA},(-1)^{\phi_{\Gamma\mu}}[D^{\mathbf{\bar{b}}}B^{\mathbf{a}}]^{\Gamma}_{\bar{\mu}}\right]\right]|\tilde{0}\rangle, \end{aligned} \tag{13}$$

5 Results and discussion

The calculations have been performed using the set of parameters shown in the table 1, which gives the values of the quark masses and strengths of the Coulomb and linear interactions and the values of the quasiparticle energy and gaps, for up and down quarks (subindex q) and, s quarks (subindex s).

$m_q[GeV]$	$m_s[GeV]$	α_C	$\beta_L[GeV^2]$	$E_q[GeV]$	$\Delta_q[GeV]$	$E_s[GeV]$	$\Delta_s[GeV]$
0.05	0.140	1.501	0.152	0.799	0.6368	0.798	0.6223

Table 1. Values of the mass parameter m_q, quasiparticle energy E_q and gap Δ_q (q = u,d quarks), m_s, E_s and Δ_s (s-quarks), α_c and β_L for Coulomb and linear parameters.

It's worth remembering that the calculations have been performed using a single set of parameters from the table 1, since the main purpose of the work is to compare the general structure of the calculated meson spectra with the experimental ones. The meson-like spectra have been calculated for the subspaces $J^P = 0^{\pm}, 1^{-}$ and $J^P = 1^{++}, 1^{+-}$, the present work only shows the case of J^P=0^- and 1^- illustrated in the Fig 1 and 2 respectively.

Fig 1 shows that the low-lying state of the pion which is quite near the experimental value, but since we have taken it as a reference state, we did not vary the parameters of the model

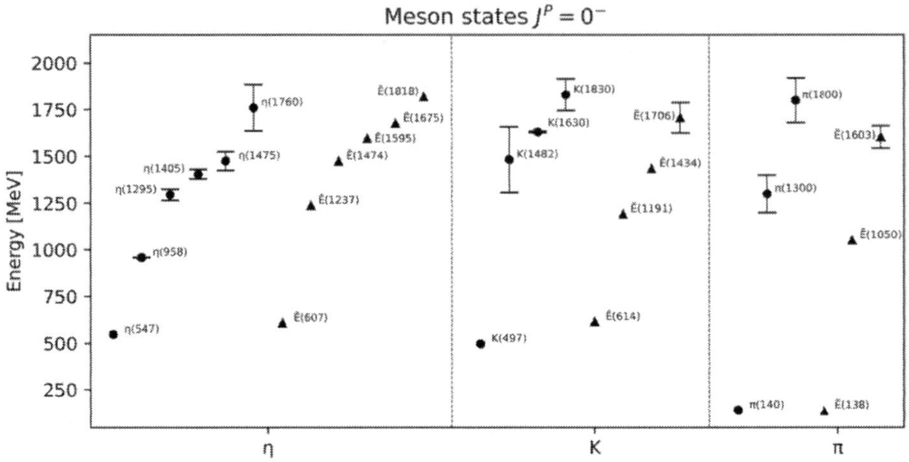

Figure 1. Energy spectrum of $J^P = 0^-$ meson states. Each of the panels shows the experimental values (dots) with the corresponding widths (vertical lines) and the calculated values (triangles). The centroids of the theoretical values are denoted by $\bar{E}$ and the vertical lines on them are results of the calculated decay widths

in order to get a better agreement with the experiment. The calculated spectrum shows two states at intermediate energy and one state at higher energy. The decay widths of these states make them compatible with data. Decay width calculation method has been worked in detail in [9, 10].

The same procedure is applied to other meson states belonging to the different subspaces considered in the calculations. In all cases, we have solved the RPA equations using the same space of quasiquark states to build the configurations restricted by the selection rules of the couplings associated to each mesonic representation.

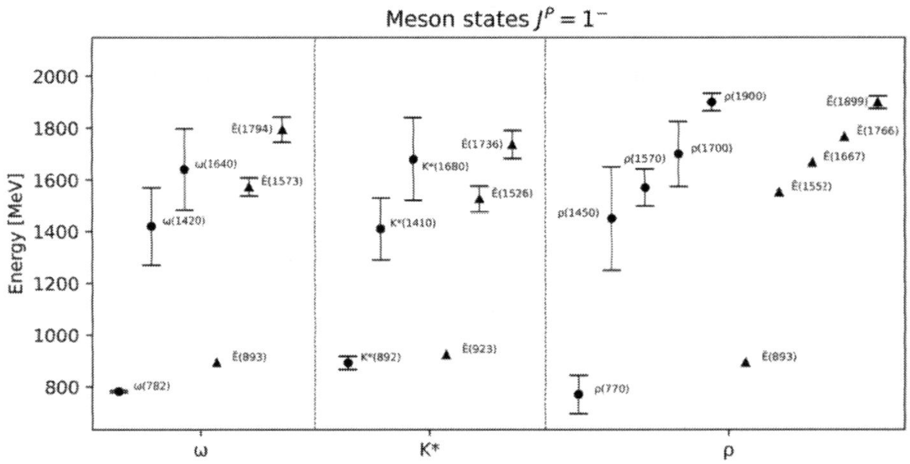

Figure 2. Experimental and theoretical results for mesons with $J^P = 1^-$. See the caption to Fig.(1)

 https://doi.org/10.1051/epjconf/202533302004

6 Conclusions

In this work, we have presented some of our results for the spectrum of meson-like states with masses up to 2 GeV. From the features exhibited by the results, it could be said that the gross structure of the observed meson spectra can be explained by the RPA method when applied to effective quasiquark degrees of freedom.

The same procedure may be also used to calculate the baryon sector. In this scheme, a three-quark state, like a single nucleon, can be described as the result of the coupling between correlated two-quasiquark states coupled to an extra quasiquark. Work is in progress about it.

References

[1] U. I. Ramirez-Soto, O. A. Rico-Trejo, T. Yépez-Martínez, P. O. Hess, A. Weber and O. Civitarese, J. Phys. G: Nucl. Part. Phys. **48**, 085013 (2021).

[2] A. P. Szczepaniak and E. S. Swanson, Phys. Rev. D **55**, 1578 (1997).

[3] A. P. Szczepaniak and E. Swanson, Phys. Rev. D **65**, 025012 (2001).

[4] D. A. Amor-Quiroz, T. Yépez-Martínez, P. O. Hess, O. Civitarese, and A. Weber, Int. J. Mod. Phys. E **26**, 1750082 (2017)

[5] T. Yépez-Martínez, P. O. Hess and O. Civitarese, Nuclear Physics A **1029**, 122566 (2023)

[6] T. D. Lee, *Particle Physics and Introduction to Field Theory* (Harwood Academic Publishers, New York, 1981).

[7] P. Ring and P. Schuck, *The Nuclear Many Body Problem* (Springer, Heidelberg, 1980).

[8] T. Yépez-Martínez, P. O. Hess, A. P. Szczepaniak and O. Civitarese, Phys. Rev. C **81**, 045204 (2010).

[9] T. Yépez-Martínez, O.A Rico-Trejo, P. O. Hess and O. Civitarese, Phys. Rev. D **110**, 114017 (2024)

[10] T. Yépez-Martínez, O. Civitarese, P. O. Hess, O. A. Rico-Trejo and U. I. Ramirez-Soto, Int. J. Mod. Phys. E **31**, 2250100 (2022)

EPJ Web of Conferences **333**, 02005 (2025) https://doi.org/10.1051/epjconf/202533302005
XLVI Symposium on Nuclear Physics 2025

Nuclear physics input to charged lepton flavor violation ($\ell \to \ell'$ conversion in nuclei)

Pablo Roig[1,2,*]

[1]Departamento de Física, Centro de Investigación y de Estudios Avanzados del Instituto Politécnico Nacional, Apartado Postal 14-740, 07000 Ciudad de México, México
[2]IFIC, Universitat de València – CSIC, Catedrático José Beltrán 2, E-46980 Paterna, Spain.

Abstract. Lepton flavor violation (LFV) has only been observed so far through neutrino oscillations, implying tiny but non-vanishing neutrino masses. The discovery of this violation in the charged lepton sector (cLFV), so far elusive, would have tremendous implications for the mechanism generating neutrino masses and possibly for understanding the universe matter-antimatter asymmetry, at energy scales prone to grand unification. In the near future, the best bounds on muon-to-electron transitions will be obtained from their conversions in nuclei, highlighting the importance of the overlap integrals between the lepton wave functions under the nucleus Coulomb potential and the nucleon densities. The electron/muon-to-tau conversion searches (which will start soon) will benefit from improved nuclear parton distribution functions (NPDFs).

1 Motivation

Neutrino masses are around six orders of magnitude smaller than the electron's. This may seem just unnatural at first -if the same Higgs mechanism [1–4] that generates the rest of the Standard Model (SM) [5–7] particle masses operates also for the neutrinos, but it is more than that.

Indeed, once a right-handed neutrino (needed for a Dirac-type mass) is introduced, a Majorana mass term becomes allowed by electroweak gauge symmetry, and this would automatically endow Majorana character (as their own antiparticles) to the light active neutrinos. This will happen unless the ultraviolet complete theory forbids lepton number violation (why then?), triggered by the Majorana neutrinos.

This type of mass-generating mechanism corresponds to the effective Weinberg operator [8] and, within the type-I seesaw mechanism [9–11], typically connects the sterile right-handed neutrinos (whose CP-violating decays generate a lepton asymmetry which would then be transferred to a baryon one, possibly explaining baryogenesis through leptogenesis [12]) to the grand unification scale ($[10^{10}, 10^{15}]$ GeV), which cannot be probed by current or planned experiments, making the traditional type-I seesaw mechanisms theoretically beautiful but phenomenologically sad.

For instance, $BR(\mu \to e\gamma) \sim \frac{\alpha}{\pi} \sum_i U^*_{ei} U_{\mu i} \frac{m^2_{\nu_i}}{M^2_W} \sim 10^{-50}$ [13–16], with an analogous GIM-like suppression [17] for $H/Z \to \ell\bar{\ell}'$, $L \to 3\ell$, $\ell \to \ell'$ in nuclei and similar processes

*e-mail: pablo.roig@cinvestav.mx

[13, 18–21]. Also, neutrinoless double beta decay searches will give an unmeasurably small signal, within the three light neutrino paradigm, in case of normal ordering of their masses (see e.g. ref. [22]).

Fortunately, low-scale seesaw mechanisms [23–25] have an additional parameter regulating the lepton number violation and can produce measurable cLFV processes, although being less susceptible to grand unification. Just as an example (whose conclusions are nevertheless quite general within this type of scenarios) among this class of models, we quote the results from ref. [26]. In this case, a pair of additional heavy O(TeV) right-handed neutrinos are added, and two massive light neutrino states can be explained (in agreement with restrictions coming from oscillation experiments, that are consistent with the lightest neutrino state being massless). This model is simple enough to easily restrict the heavy-light mixing driving the LFV processes. Within it, in the $\mu - e$ sector, the current strongest constraints come from the $\mu \to e\gamma$ MEG bound [27] (branching ratio, BR, $< 4.2 \times 10^{-13}$ at 90% C.L.). However, the improvement in the plane of the $e - \mu$ heavy-light mixings by the MEG-II [28] results (up to BR $< 6 \times 10^{-14}$, at 90% C.L.) will be superseded by the $\mu \to e$ conversion in nuclei, that can reach up to the 10^{-18} exclusion level in COMET at J-PARC [29] and Mu2e at FNAL [30]. Model-independently, the combined constraints on the $\ell \to \ell'$ cLFV transitions have been worked out minutely in an effective field theory framework for heavy new physics (see for instance refs. [31–40]) and the current best constraint on the underlying new physics scale is at the level of $\Lambda \gtrsim 10^5$ TeV for the electromagnetic dipole operator mediating $\mu \to e$ transitions. $\mu/e \to \tau$ conversion in nuclei has recently been revived [34, 41–46] and will be studied at future facilities, such as NA62, EIC, ILC, LHeC, etc. However, it is easily seen, that it will only compete with the bounds from $\mu \to e$ transitions if the cLFV couplings are proportional to the daughter lepton mass.

We will shortly review in the next section the nuclear physics inputs needed to interpret the results of the searches for $\ell \to \ell'$ conversions in nuclei, whose relevance was just highlighted.

2 Nuclear physics inputs

A classical reference for the discussion of $\mu \to e$ conversions in nuclei is [47], to which we refer the reader for further details.

The effective Lagrangian for $\mu \to e$ conversions in nuclei contains dipole operators (giving rise to $\ell \to \ell'\gamma$ decays at tree level) and four-fermion structures, composed of a $\ell - \ell'$ current and a quark one, with all possible Dirac matrices in both bilinears, that is, (pseudo)scalar, (pseudo)vector and tensor currents. The initial wavefunction corresponds to the $1s$ configuration of the muonic atom, with binding energy ϵ_b, and the final one is an eigenstate with energy $m_\mu - \epsilon_b$. Both wavefunctions are determined by solving the Dirac equation in presence of the electric field of the nucleus, E. We are generally mostly interested in the dominant [48] coherent case, where the initial and final states are identical. Thus, the only non-vanishing relevant matrix elements are $\langle N|\bar{q}q|N\rangle$ and $\left\langle N\middle|\bar{q}\gamma^0 q\middle|N\right\rangle$, which are straightforwardly related to the proton and neutron densities in nuclei, $\rho^{(p/n)}$:

$$\langle N|\bar{q}q|N\rangle = ZG_S^{(q,p)}\rho^{(p)} + (A-Z)G_S^{(q,n)}\rho^{(n)}\,, \quad \left\langle N\middle|\bar{q}\gamma^0 q\middle|N\right\rangle = 2Z\rho^{(p)} + (A-Z)\rho^{(n)}, \qquad (1)$$

for $q = u$, with $Z \leftrightarrow (A-Z)$ for $q = d$. The $G_S^{(q,p/n)}$ are $O(5)$, according to ref. [49].

The conversion probability depends on the effective couplings $G_S^{(q,p/n)}$ and the overlap integrals D, $(S/V)^{(p,n)}$, which are functions, in turn, on the muon and electron wavefunctions (Ψ), the nucleon densities $\rho^{(p,n)}$ (determining the electric field in the nucleus, that is also needed) and the mass and atomic numbers, A and Z, respectively. For instance, the conversion probability normalized to $2G_F^2$ is given by a sum of contributions from the corresponding overlap integral times a Lagrangian coupling. While D integrates the nucleus electric field, S and V integrate the nucleon densities times wavefunctions weighted by Z(A-Z) for protons(neutrons).

We will give just an example contribution to understand how this works; the rest can be checked e. g. in ref. [47]. The Lagrangian includes the term (P_R is the right-handed chiral projector)

$$\mathcal{L} = \cdots - \frac{G_F}{\sqrt{2}} \sum_{q=u,d,s} (g_{LS}(q)\bar{e}P_R\mu + \ldots)\bar{q}q + \ldots, \tag{2}$$

yielding the probability amplitude (usual conventions for the wavefunction quantum numbers are followed)

$$\mathcal{M} = \cdots + \frac{G_F}{\sqrt{2}} \sum_{q=u,d,s} \int d^3x \left(g_{LS}(q)\bar{\Psi}^{\mu(e)}_{\kappa,W} P_R \Psi^{(\mu)}_{1S} + \ldots \right) \langle N'|\bar{q}q|N\rangle + \ldots \tag{3}$$

The normalized conversion probability is then $\widetilde{\omega}_{conv} = |\cdots + \widetilde{g}_{LS}(p)S^{(p)} + \ldots|^2$, where $\widetilde{g}_{LS}(p) = \sum G_S^{(q,p)} g_{LS}(q)$, with

$$S^{(p)} = \frac{1}{2\sqrt{2}} \int_0^\infty drr^2 Z\rho^{(p)} \left(g_e^-(r)g_\mu^-(r) - f_e^-(r)f_\mu^-(r) \right). \tag{4}$$

It is interesting to note that, in the massless electron limit, only two radial wavefunctions are independent, $g_e^+ = if_e^-$ and $if_e^+ = g_e^-$. For the convenience of the reader we quote the other overlap integrals ($Z \to (A-Z)$ for neutrons) in the following:

$$D = \frac{4}{\sqrt{2}} m_\mu \int_0^\infty drr^2[-E(r)](g_e^- f_\mu^- + f_e^- g_\mu^-), \quad V^{(p)} = \frac{1}{2\sqrt{2}} \int_0^\infty drr^2 Z\rho^{(p)}(g_e^- g_\mu^- + f_e^- f_\mu^-). \tag{5}$$

Proton densities are extracted model-independently from $e-p$ scattering experiments. Neutron densities are much more difficult, in contrast. Information about them comes from pionic atoms, scattering experiments with protons, α particles or pions, as well as experiments with anti-proton atoms. For light nuclei, assuming $\rho^{(n)} = \rho^{(p)}$ works at the percent level. For heavy nuclei it is better to use π-atoms data or polarized proton scattering measurements, although the former tend to give $10-20\%$ smaller overlap integrals. In any case, it is clear that conversion rates increase with Z for $Z \leq 30$, are highest for $30 \leq Z \leq 60$, and then decrease. In general, uncertainties are concluded to be relatively small for Ti ($Z = 22$), but not for Au ($Z = 79$) or Pb ($Z = 82$). Ab initio calculations of these integrals (for ^{27}Al, ^{48}Ca and ^{48}Ti) with improved determinations of charge distributions from electron scattering were recently presented in ref. [50], including full covariance matrices, representing a significant step forward in reducing nuclear structure uncertainties for the new physics interpretation of the $\mu \to e$ conversion in nuclei searches. It is worth noting that the MuonBridge code [39] is available for systematic EFT studies of these processes.

The $\ell \to \tau$ ($\ell = e,\mu$) conversions in nuclei searches are done in deep inelastic scattering experiments, with an $O(100)$ GeV ℓ beam hitting a nucleus in fixed targets of Fe, Pb, etc. In this case the parton level process can be described in perturbative QCD and the rest depends on the nuclear PDFs. Those of the nCTEQ15 project [51] are incorporated in the Mathematica package ManeParse [52]. In the quoted references it is seen that, until a ratio between the conversion and capture rates of $\sim 10^{-15}$ is reached (it is $\sim$ 100 times larger at NA-64 [42]), these processes will not compete with the bounds on τ cLFV coming from BaBar or Belle [53] (improved at Belle-II [54]).

3 Conclusions and outlook

The generation of neutrino masses generalizes and extends necessarily the SM (either through the effective Weinberg operator, or through its non-realization, which would point to fundamental lepton number conservation in the ultraviolet). In the most beautiful theoretical realization, it will be linked to both grand unification and baryogenesis through leptogenesis. Unfortunately, not only lepton flavor, but also lepton number violation (specifically neutrinoless double beta decay) may escape detection, forbidding its experimental verification. Low-scale seesaws have been proposed as alternatives. These are less simple and elegant models, without straightforward connection to grand unified theories but yielding O(TeV) states potentially observable at present/near-future colliders, with cLFV signatures in the reach of current/forthcoming facilities. Within them, $\mu \to e$ conversion in nuclei will be giving the strongest limits thanks to FNAL and J-PARC experiments in a few years. The interplay of these with $\mu \to 3e, \mu \to e\gamma, \ldots$ will be key in identifying the underlying new physics. This task will be cleaner if the uncertainties in the overlap integrals are reduced, which depends critically on nuclear physics input. The tau sector could in principle be more sensitive to new physics, provided this is mass-dependent. Conversions in nuclei depend in this case on NPDFs, but will need better sensitivity than the one projected for NA64 by $\sim$ two orders of magnitude, in order to reach the level of other cLFV processes involving the τ flavor.

Acknowledgements

Besides the excellent organizers work, P. Roig acknowledges Mexican support from CONAHCYT, particularly from project CBF2023-2024-3226, as well as Spanish support during his sabbatical through projects MCIN/AEI/10.13039/501100011033, grants PID2020-114473GB-I00 and PID2023-146220NB-I00, and Generalitat Valenciana grant PROMETEO/2021/07.

References

[1] P. W. Higgs, "Broken symmetries, massless particles and gauge fields," Phys. Lett. **12** (1964), 132-133. https://doi.org/10.1016/0031-9163(64)91136-9

[2] F. Englert and R. Brout, "Broken Symmetry and the Mass of Gauge Vector Mesons," Phys. Rev. Lett. **13** (1964), 321-323. https://doi.org/10.1103/PhysRevLett.13.321

[3] P. W. Higgs, "Broken Symmetries and the Masses of Gauge Bosons," Phys. Rev. Lett. **13** (1964), 508-509. https://doi.org/10.1103/PhysRevLett.13.508

[4] G. S. Guralnik, C. R. Hagen and T. W. B. Kibble, "Global Conservation Laws and Massless Particles," Phys. Rev. Lett. **13** (1964), 585-587. https://doi.org/10.1103/PhysRevLett.13.585

[5] S. L. Glashow, "Partial Symmetries of Weak Interactions," Nucl. Phys. **22** (1961), 579-588. https://doi.org/10.1016/0029-5582(61)90469-2

[6] S. Weinberg, "A Model of Leptons," Phys. Rev. Lett. **19** (1967), 1264-1266. https://doi.org/10.1103/PhysRevLett.19.1264

[7] A. Salam, "Weak and Electromagnetic Interactions," Conf. Proc. C **680519** (1968), 367-377. https://doi.org/10.1142/9789812795915_0034

[8] S. Weinberg, "Baryon and Lepton Nonconserving Processes," Phys. Rev. Lett. **43** (1979), 1566-1570. https://doi.org/10.1103/PhysRevLett.43.1566

[9] P. Minkowski, "$\mu \to e\gamma$ at a Rate of One Out of 10^9 Muon Decays?," Phys. Lett. B **67** (1977), 421-428. https://doi.org/10.1016/0370-2693(77)90435-X

[10] M. Gell-Mann, P. Ramond and R. Slansky, "Complex Spinors and Unified Theories," Conf. Proc. C **790927** (1979), 315-321.

[11] R. N. Mohapatra and G. Senjanovic, "Neutrino Mass and Spontaneous Parity Nonconservation," Phys. Rev. Lett. **44** (1980), 912. https://doi.org/10.1103/PhysRevLett.44.912

[12] M. Fukugita and T. Yanagida, "Baryogenesis Without Grand Unification," Phys. Lett. B **174** (1986), 45-47. https://doi.org/10.1016/0370-2693(86)91126-3

[13] S. T. Petcov, "The Processes $\mu \to e + \gamma, \mu \to e + \bar{e}, \nu' \to \nu + \gamma$ in the Weinberg-Salam Model with Neutrino Mixing," Sov. J. Nucl. Phys. **25** (1977), 340 [erratum: Sov. J. Nucl. Phys. **25** (1977), 698; erratum: Yad. Fiz. **25** (1977), 1336] JINR-E2-10176.

[14] T. P. Cheng and L. F. Li, "Nonconservation of Separate mu - Lepton and e - Lepton Numbers in Gauge Theories with v+a Currents," Phys. Rev. Lett. **38** (1977), 381. https://doi.org/10.1103/PhysRevLett.38.381

[15] W. J. Marciano and A. I. Sanda, "Exotic Decays of the Muon and Heavy Leptons in Gauge Theories," Phys. Lett. B **67** (1977), 303-305. https://doi.org/10.1016/0370-2693(77)90377-X

[16] B. W. Lee and R. E. Shrock, "Natural Suppression of Symmetry Violation in Gauge Theories: Muon - Lepton and Electron Lepton Number Nonconservation," Phys. Rev. D **16** (1977), 1444. https://doi.org/10.1103/PhysRevD.16.1444.

[17] S. L. Glashow, J. Iliopoulos and L. Maiani, "Weak Interactions with Lepton-Hadron Symmetry," Phys. Rev. D **2** (1970), 1285-1292. https://doi.org/10.1103/PhysRevD.2.1285

[18] J. I. Illana and T. Riemann, "Charged lepton flavor violation from massive neutrinos in Z decays," Phys. Rev. D **63** (2001), 053004. https://doi.org/10.1103/PhysRevD.63.053004

[19] E. Arganda, A. M. Curiel, M. J. Herrero and D. Temes, "Lepton flavor violating Higgs boson decays from massive seesaw neutrinos," Phys. Rev. D **71** (2005), 035011. https://doi.org/10.1103/PhysRevD.71.035011

[20] G. Hernández-Tomé, G. López Castro and P. Roig, "Flavor violating leptonic decays of τ and μ leptons in the Standard Model with massive neutrinos," Eur. Phys. J. C **79** (2019) no.1, 84 [erratum: Eur. Phys. J. C **80** (2020) no.5, 438]. https://doi.org/10.1140/epjc/s10052-019-6563-4

[21] P. Blackstone, M. Fael and E. Passemar, "$\tau \to \mu\mu\mu$ at a rate of one out of 10^{14} tau decays?," Eur. Phys. J. C **80** (2020) no.6, 506. https://doi.org/10.1140/epjc/s10052-020-8059-7

[22] J. D. Vergados, H. Ejiri and F. Simkovic, "Theory of Neutrinoless Double Beta Decay," Rept. Prog. Phys. **75** (2012), 106301. https://doi.org/10.1088/0034-4885/75/10/106301

[23] R. N. Mohapatra and J. W. F. Valle, "Neutrino Mass and Baryon Number Nonconservation in Superstring Models," Phys. Rev. D **34** (1986), 1642. https://doi.org/10.1103/PhysRevD.34.1642

[24] J. Bernabéu, A. Santamaría, J. Vidal, A. Méndez and J. W. F. Valle, "Lepton Flavor Nonconservation at High-Energies in a Superstring Inspired Standard Model," Phys. Lett. B **187** (1987), 303-308. https://doi.org/10.1016/0370-2693(87)91100-2

[25] M. Malinsky, J. C. Romao and J. W. F. Valle, "Novel supersymmetric SO(10) seesaw mechanism," Phys. Rev. Lett. **95** (2005), 161801. https://doi.org/10.1103/PhysRevLett.95.

161801

[26] G. Hernández-Tomé, J. I. Illana, M. Masip, G. López Castro and P. Roig, "Effects of heavy Majorana neutrinos on lepton flavor violating processes," Phys. Rev. D **101** (2020) no.7, 075020, https://doi.org/10.1103/PhysRevD.101.075020.

[27] A. M. Baldini *et al.* [MEG], "Search for the lepton flavour violating decay $\mu^+ \to e^+\gamma$ with the full dataset of the MEG experiment," Eur. Phys. J. C **76** (2016) no.8, 434. https://doi.org/10.1140/epjc/s10052-016-4271-x

[28] A. M. Baldini *et al.* [MEG II], "The design of the MEG II experiment," Eur. Phys. J. C **78** (2018) no.5, 380. https://doi.org/10.1140/epjc/s10052-018-5845-6

[29] R. Abramishvili *et al.* [COMET], "COMET Phase-I Technical Design Report," PTEP **2020** (2020) no.3, 033C01. https://doi.org/10.1093/ptep/ptz125

[30] L. Bartoszek *et al.* [Mu2e], "Mu2e Technical Design Report," https://doi.org/10.2172/1172555

[31] A. Crivellin, S. Davidson, G. M. Pruna and A. Signer, "Renormalisation-group improved analysis of $\mu \to e$ processes in a systematic effective-field-theory (EFT) approach," JHEP **05** (2017), 117. 10.1007/JHEP05(2017)117

[32] A. Crivellin, S. Najjari and J. Rosiek, "Lepton Flavor Violation in the Standard Model with general Dimension-Six Operators," JHEP **04** (2014), 167. 10.1007/JHEP04(2014)167

[33] V. Cirigliano, S. Davidson and Y. Kuno, "Spin-dependent $\mu \to e$ conversion," Phys. Lett. B **771** (2017), 242-246. https://doi.org/10.1016/j.physletb.2017.05.053

[34] T. Husek, K. Monsálvez-Pozo and J. Portolés, "Lepton-flavour violation in hadronic tau decays and $\mu - \tau$ conversion in nuclei," JHEP **01** (2021), 059. https://doi.org/10.1007/JHEP01(2021)059

[35] S. Banerjee, *et al.* "Snowmass 2021 White Paper: Charged lepton flavor violation in the tau sector," [arXiv:2203.14919 [hep-ph]].

[36] S. Davidson and B. Echenard, "Reach and complementarity of $\mu \to e$ searches," Eur. Phys. J. C **82** (2022) no.9, 836. https://doi.org/10.1016/10.1140/epjc/s10052-022-10773-4

[37] M. Hoferichter, J. Menéndez and F. Noël, "Improved Limits on Lepton-Flavor-Violating Decays of Light Pseudoscalars via Spin-Dependent $\mu \to e$ Conversion in Nuclei," Phys. Rev. Lett. **130** (2023) no.13, 131902. https://doi.org/10.1103/PhysRevLett.130.131902

[38] E. Fernández-Martínez, X. Marcano and D. Naredo-Tuero, "Global lepton flavour violating constraints on new physics," Eur. Phys. J. C **84** (2024) no.7, 666. https://doi.org/10.1016/10.1140/epjc/s10052-024-12973-6

[39] W. Haxton, K. McElvain, T. Menzo, E. Rule and J. Zupan, "Effective theory tower for $\mu \to e$ conversion," JHEP **11** (2024), 076. https://doi.org/10.1007/JHEP11(2024)076

[40] F. Delzanno, K. Fuyuto, S. Gonzàlez-Solís and E. Mereghetti, "Global analysis of $\mu \to e$ interactions in the SMEFT," [arXiv:2411.13497 [hep-ph]].

[41] S. N. Gninenko, M. M. Kirsanov, N. V. Krasnikov and V. A. Matveev, "Probing lepton flavor violation in muon-neutrino + N -> tau + ... scattering and mu -> tau conversion on nucleons," Mod. Phys. Lett. A **17** (2002), 1407. https://doi.org/10.1142/S0217732302007855

[42] S. Gninenko, S. Kovalenko, S. Kuleshov, V. E. Lyubovitskij and A. S. Zhevlakov, "Deep inelastic $e - \tau$ and $\mu - \tau$ conversion in the NA64 experiment at the CERN SPS," Phys. Rev. D **98** (2018) no.1, 015007. https://doi.org/10.1103/PhysRevD.98.015007

[43] I. Pacheco and P. Roig, "Lepton flavor violation in the Littlest Higgs Model with T parity realizing an inverse seesaw," JHEP **02** (2022), 054. https://doi.org/10.1007/JHEP02(2022)054

[44] T. Husek, K. Monsálvez-Pozo and J. Portolés, "Constraints on leptoquarks from lepton-flavour-violating tau-lepton processes," JHEP **04** (2022), 165. https://doi.org/10.1007/10.1007/JHEP04(2022)165

[45] E. Ramírez and P. Roig, "Lepton flavor violation within the simplest little Higgs model," Phys. Rev. D **106** (2022) no.5, 056018. https://doi.org/10.1103/PhysRevD.106.056018

[46] F. Fortuna, X. Marcano, M. Marín and P. Roig, "Lepton flavor violation from diphoton effective interactions," Phys. Rev. D **108** (2023) no.1, 015008. https://doi.org/10.1103/PhysRevD.108.015008

[47] R. Kitano, M. Koike and Y. Okada, "Detailed calculation of lepton flavor violating muon electron conversion rate for various nuclei," Phys. Rev. D **66** (2002), 096002 [erratum: Phys. Rev. D **76** (2007), 059902]. https://doi.org/10.1103/PhysRevD.76.059902

[48] S. Weinberg and G. Feinberg, "Electromagnetic Transitions Between mu Meson and Electron," Phys. Rev. Lett. **3** (1959), 111-114. https://doi.org/10.1103/PhysRevLett.3.111

[49] T. S. Kosmas, G. K. Leontaris and J. D. Vergados, "Lepton flavor nonconservation," Prog. Part. Nucl. Phys. **33** (1994), 397-448. https://doi.org/10.1016/0146-6410(94)90047-7

[50] M. Heinz, M. Hoferichter, T. Miyagi, F. Nöel and A. Schwenk, "Ab initio calculations of overlap integrals for $\mu \to e$ conversion in nuclei," [arXiv:2412.04545 [nucl-th]].

[51] K. Kovarik, A. Kusina, T. Jezo, D. B. Clark, C. Keppel, F. Lyonnet, J. G. Morfin, F. I. Olness, J. F. Owens and I. Schienbein, *et al.* "nCTEQ15 - Global analysis of nuclear parton distributions with uncertainties in the CTEQ framework," Phys. Rev. D **93** (2016) no.8, 085037. https://doi.org/10.1103/PhysRevD.93.085037

[52] D. B. Clark, E. Godat and F. I. Olness, "ManeParse : A Mathematica reader for Parton Distribution Functions," Comput. Phys. Commun. **216** (2017), 126-137. https://doi.org/10.1016/j.cpc.2017.03.004

[53] A. J. Bevan *et al.* [BaBar and Belle], "The Physics of the B Factories," Eur. Phys. J. C **74** (2014), 3026. https://doi.org/10.1140/epjc/s10052-014-3026-9

[54] E. Kou *et al.* [Belle-II], "The Belle II Physics Book," PTEP **2019** (2019) no.12, 123C01 [erratum: PTEP **2020** (2020) no.2, 029201]. https://doi.org/10.1093/ptep/ptz106

Optimization of the Magnetic Environment Inside a Magnetically Shielded Enclosure for the Measurement of the Neutron Electric Dipole Moment

Erika Ruíz[1,*], *Libertad* Barrón-Palos[1,**], and *the nEDM@SNS collaboration*

[1]Instituto de Física, Universidad Nacional Autónoma de México, Apartado Postal 20-364, 01000, Mexico

Abstract. The search of the Electric Dipole Moment of the neutron (nEDM) is the current goal of several experiments around the world. One key element for these experiments is a very stringent control of the magnetic field environment in which the nEDM is measured. In the context of an experiment proposed for the Spallation Neutron Source of the Oak Ridge National Laboratory (nEDM@SNS), we developed a COMSOL® simulation of the Magnetically Shielded Enclosure (MSE) of the experiment, as well as an optimization of the Earth's Magnetic Field Compensation System surrounding the MSE that minimized the magnetic field gradients in the Measurement Region.

1 Introduction

A non-zero neutron Electric Dipole Moment (nEDM) would violate time reversal invariance T, or CP under the assumption of CPT invariance. In such a case, the nEDM, $\vec{d}_n$, would couple to an external electric field E analogous to the neutron magnetic moment, $\vec{\mu}_n$, coupling to an external magnetic field B, as expressed in the Hamiltonian:

$$H = -(\vec{\mu}_n \cdot \vec{B} + \vec{d}_n \cdot \vec{E}). \tag{1}$$

The experimental approach to measure the nEDM has traditionally been Ramsey's method of separated oscillatory fields [1], in which the neutron's frequency of precession is measured in the presence of external electric and magnetic fields (perpendicular to the neutron spin direction $\vec{\sigma}_n$), in configurations where these fields are either parallel or antiparallel. The difference in precession frequency under these conditions is directly proportional to $d_n = \vec{d}_n \cdot \vec{\sigma}_n$:

$$d_n = \frac{\hbar(\omega_{\downarrow\uparrow} - \omega_{\uparrow\uparrow})}{4E}. \tag{2}$$

The current experimental limit for the nEDM is $|d_n| < 1.8 \times 10^{-26}\, e\,\mathrm{cm}$ (90% C.L.) [2]. Given the typical electric field strengths used in nEDM experiments, one can appreciate the extremely small frequency shift between the two studied configurations of external electromagnetic fields, $\omega_{\downarrow\uparrow} - \omega_{\uparrow\uparrow}$. For an electric field of 10 kV/cm, this shift is expected to be

*e-mail: erikaruiz@estudiantes.fisica.unam.mx
**e-mail: libertad@fisica.unam.mx

on the order of 1 μHz at current limit on d_n. In the presence of an external magnetic field, the neutron precession frequency around this field, $\omega_L = \gamma_n B$ (Larmor frequency [3], where γ_n is the neutron gyromagnetic ratio), will change by more than 1 μHz if the field amplitude fluctuates by as little as 10 fT. Thus, a very stringent control of the magnetic environment is an essential requirement for any experiment aiming to measure the nEDM.

A key element in the nEDM@SNS experiment[1] was the development of a Magnetically Shielded Enclosure (MSE), a rectangular prism-shaped room, consisting of two mu-metal layers [3]. To improve the performance of the MSE, a set of electromagnets surrounding its exterior, herein referred to as the compensation system, was designed to produce a magnetic field that would partially cancel the earth's magnetic field on the external surface of the MSE. To achieve the precision goal of the experiment ($\delta d_n < 2\text{-}3 \times 10^{-28}\, e$cm), magnetic fields within a 1 m^3 cube encompassing the measurement region inside the MSE were required to remain below 10 nT. More critically, magnetic field gradients in the same region needed to stay below 1 nT/m [3]. The finite element simulation using COMSOL Multiphysics®, developed to analyze the performance of the MSE and the compensation system, is described in Section 2. The steps taken to optimize the original electric current design for the compensation system, along with the results of this optimization, are presented in Section 3. Finally, Section 4 summarizes the conclusions of this work.

2 Simulation of the MSE and Compensation System

To study the performance of the MSE and the compensation system, we developed a simulation using the AC/DC module of COMSOL Multiphysics®.

The MSE is a closed room in the shape of a rectangular prism, composed of two layers of 3 mm thick mu-metal, separated by 0.4 m. Its inner dimension are 4.1 m × 4.1 m × 6.1 m. It was intended to be located at the end of the FnPB BL13-A beamline of the SNS, which would deliver polarized ultracold neutrons (UCNs) to its interior. The orientation of the beamline, and therefore that of the MSE, is 39.79° with respect to the geographic North. The earth's magnetic field at this location was obtained from the National Oceanic and Atmospheric Administration (NOAA) calculator [4], specifically from the World Magnetic Model 2020. The values used in our simulation correspond to those expected for November 2024; −44.449 μT for the vertical field, 21.959 μT for the north component, and −2.252 μT for the east one. To compensate for this magnetic field on the external surface of the MSE, where magnetic fields no larger than 30 μT were required, a system consisting of three square coils, co-centered with the MSE and distributed along the vertical axis, and one diagonal coil, was conceived; we will refer to these as vertical and diagonal coils. The arrangement is shown in Figure 1. The electric current of these coils in the original design was 120 A· turns for each one of the three vertical coils and 100 A· turns for the diagonal coil.

The walls of the MSE feature various holes intended to serve as feedthroughs for different instruments, with the largest being the hole for the neutron guide access. The MSE is also shown in the schematic in Figure 1, along with a small cube (1 m^3) included in the model, which contains the measurement region where a finer mesh (higher resolution) was used in the simulation.

The material properties used in the simulation were taken from the COMSOL Multiphysics Material Library. Specifically, MuMETAL® was used for the MSE walls, with relative permeability defined by its corresponding B-H curve; copper was used for the coils of

[1]The nEDM@SNS experiment, originally proposed and partially developed at the Spallation Neutron Source (SNS) of the Oak Ridge National Laboratory, was cancelled in 2023. The unique features of this experiment, based on the approach proposed in [1] and expected to improve the experimental limit on the nEDM by two orders of magnitude, are now being pursued by the nEDMSF initiative [5].

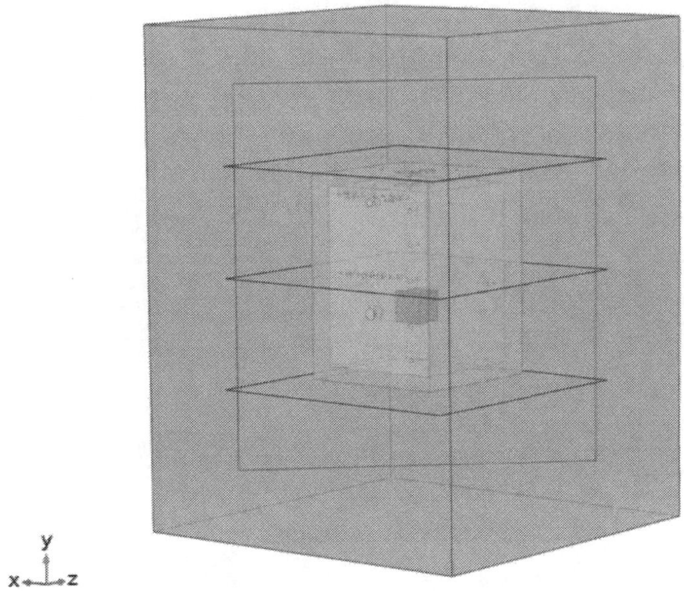

Figure 1. Schematic representation of the MSE and compensation system, consisting of three square coils distributed along the vertical axis (blue lines) and one diagonal coil (red line). The small cube (1 m^3) inside the MSE highlights the measurement region. The outermost prism represents the boundary of the simulated region.

the compensation system; and air was assigned to both the ambient prism (the boundary of the simulated region) and the cube representing the measurement region.

Using the Magnetic Fields, No Currents (mfnc) module, we simulated the earth's magnetic field using the Magnetic Flux Density option. For the compensation system coils, we used the Magnetic Fields (mf) module, setting the currents for the diagonal coil independent from that of the three vertical coils. In order to be able to compute derivatives of the simulated magnetic fields, and therefore study field gradients, we used the Coefficient Form PDE module. Only stationary studies were conducted.

3 Optimization of the Compensation System

The total magnetic field

$$\vec{B}_T = \vec{B}_E + \vec{B}_D + \vec{B}_V, \tag{3}$$

where B_E refers to the earth's magnetic field and B_V and B_D denote the fields produced by the Vertical and Diagonal Coils, respectively, was computed for different combinations of N_V and N_D values. The original compensation system had 120 A · turns current in the vertical coils and a 100 A · turns current in the diagonal coil, corresponding to $N_V^{\text{orig}} = 120$ and $N_D^{\text{orig}} = 100$. The average total gradient in the measurement region, defined as

$$G = \sqrt{\sum_{i=1}^{3}(L_i + T_i)^2}, \tag{4}$$

where $L_i = \langle \partial_i B_{T,i} \rangle$ and $T_i = \langle \partial_i (B_{T,j} + B_{T,k}) \rangle$, with $i \neq j \neq k$, represent the average linear and transverse components of the gradient in the i-th direction, as shown in Figure 2. A minimum in the total gradient is observed for what we define as the optimal values, $N_V^{\text{opt}} = 163$ and $N_D^{\text{opt}} = 294$.

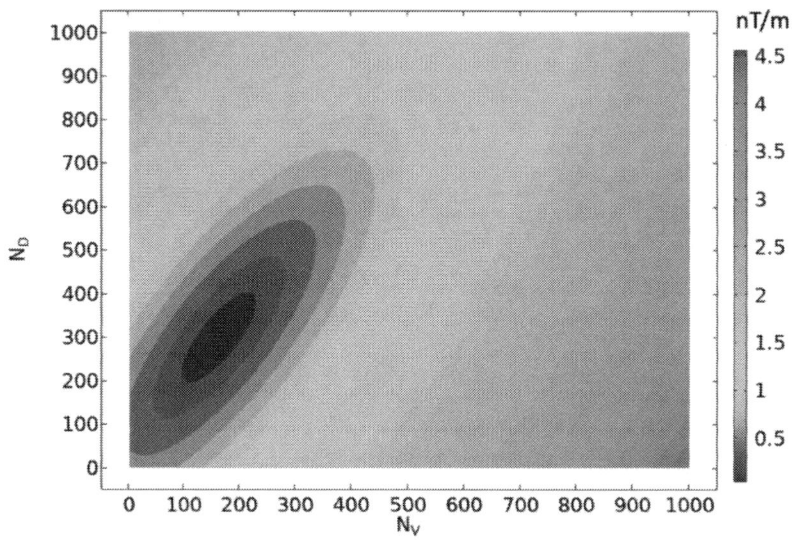

Figure 2. Average total gradient in the measurement region as a function of N_V and N_D. The optimal values, $N_V^{\text{opt}} = 163$ and $N_D^{\text{opt}} = 294$, were determined from this analysis.

3.1 Bulk Magnetic Fields

A comparison of the magnetic fields achieved inside the MSE, as well as on its external surface, is presented in Figures 3 and 4 for three cases: *i*) No compensation system; *ii*) Compensation system with the original design electric currents (120 A· turns for vertical coils and 100 A· turns for diagonal coil); and *iii*) Optimized compensation system (163 A· turns for vertical coils and 294 A· turns for diagonal coil). It can be seen that the quality of the magnetic environment inside the MSE, as well as the magnetic field values on its external surface, are significantly reduced by compensating the ambient fields. However, it is important to point out that both configurations of the compensation system meet the requirement of maintaining the total magnetic field below 10 nT in the measurement region, averaging 7.68 nT for the original and 3.75 nT for the optimized setup. Similarly, the magnetic field on the external surface of the MSE remains below 30 μT in both cases, with average values of 16.22 μT for the original and 11.40 μT for the optimized configuration.

3.2 Magnetic Field Gradients and Uniformity

Figure 5 shows the three components of the transverse gradients in the measurement region for the original compensation system design (top) and the optimized configuration (bottom).

Figure 3. Total magnetic field in the central plane of the MSE. From left to right: *i*) No compensation system; *ii*) Compensation system with the original design electric currents, and *iii*) Optimized compensation system. The direction and amplitude of the ambient magnetic field is also indicated by green arrow vectors.

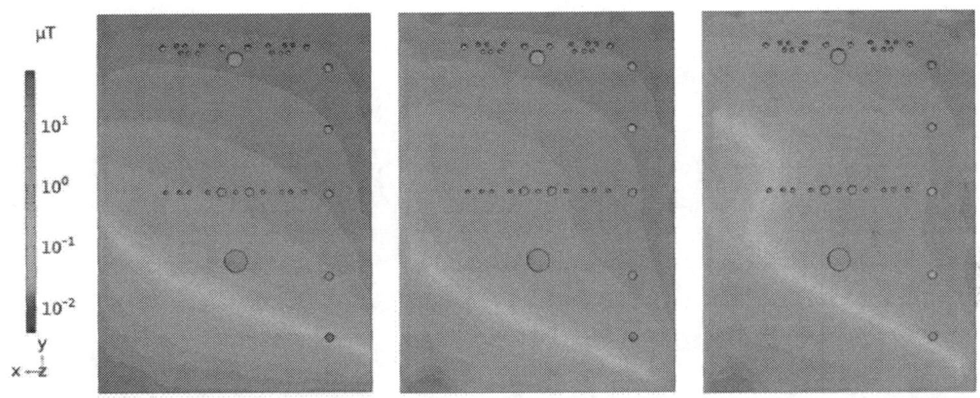

Figure 4. Example of the total magnetic field on one of the outer surfaces of the MSE. From left to right: *i*) No compensation system; *ii*) Compensation system with the original design electric currents, and *iii*) Optimized compensation system.

A similar behavior is observed for the linear gradients, although their values are generally smaller. In both configurations, the gradients remain below 1 nT/m throughout most of the 1 m^3 volume encompassing the measurement region. However, the optimized case achieves a noticeable reduction in both the overall gradient values and their spatial variability, highlighting the benefits of optimizing the compensation system.

The uniformity of the magnetic field inside the MSE, particularly within the measurement region can be evaluated through the residual, defined as

$$\text{Residual} = \left| \frac{B_T(x, y, z) - \langle B_T \rangle}{\langle B_T \rangle} \right| 100\%, \tag{5}$$

where $\langle B_T \rangle$ represents the average total field in the measurement region and $B_T(x, y, z)$ the total magnetic field value in a particular point within the same region. Figure 6 shows the

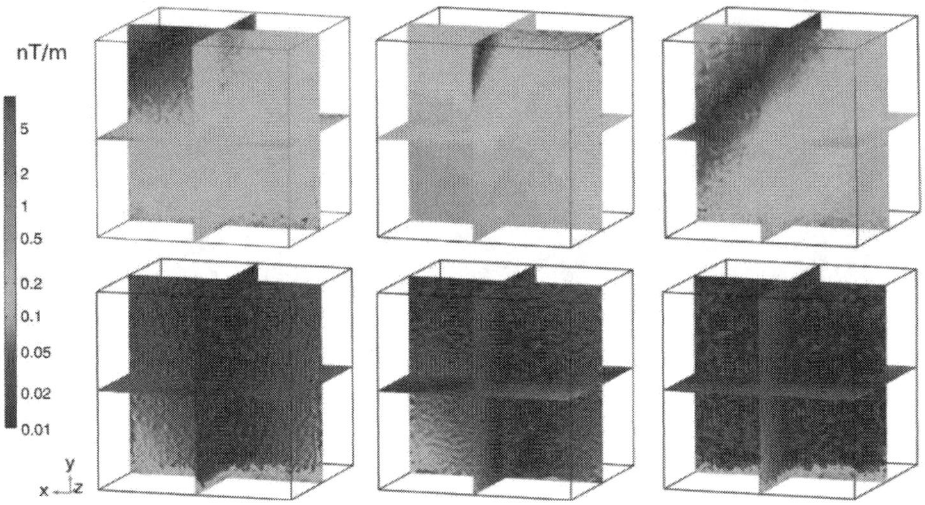

Figure 5. Components of the transverse magnetic field gradient (from left to right, T_x, T_y and T_z) for the original (top) and optimized (bottom) compensation system.

Residual values inside the measurement region for the original (left) and optimized (right) compensation system. A significantly more uniform field is achieved in the optimized case, with a percentage variation below 1.58 %, compared to up to 2.72 % in the original system.

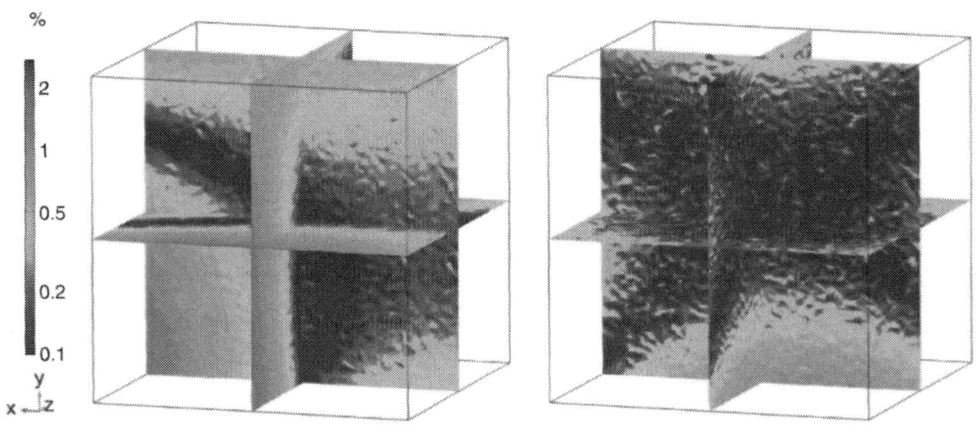

Figure 6. Magnetic field uniformity in the measurement region analyzed as the residual (see Equation 5) for the original (left) and optimized (right) compensation system.

3.3 Shielding Factor

Another crucial aspect to characterize the optimal MSE + compensation system arrangement is the shielding factor, defined as the ratio between the earth's magnetic field (assuming the

the MSE and compensation were not present) and the actual magnetic field obtained in the simulation when these elements are included. Figure 7 presents the shielding factor along the axis defined by the center of the neutron guide for: *i*) No compensation system, *ii*) Compensation system with the original design electric currents, and *iii*) Optimized compensation system. Inside the MSE, an increase in the shielding factor by a factor of 2 is observed when comparing the original and optimized compensation systems. The asymmetry observed in Figure 7 between the upstream and downstream ends of the MSE reflects the presence of the neutron guide access hole on the upstream side, which is absent on the opposite end.

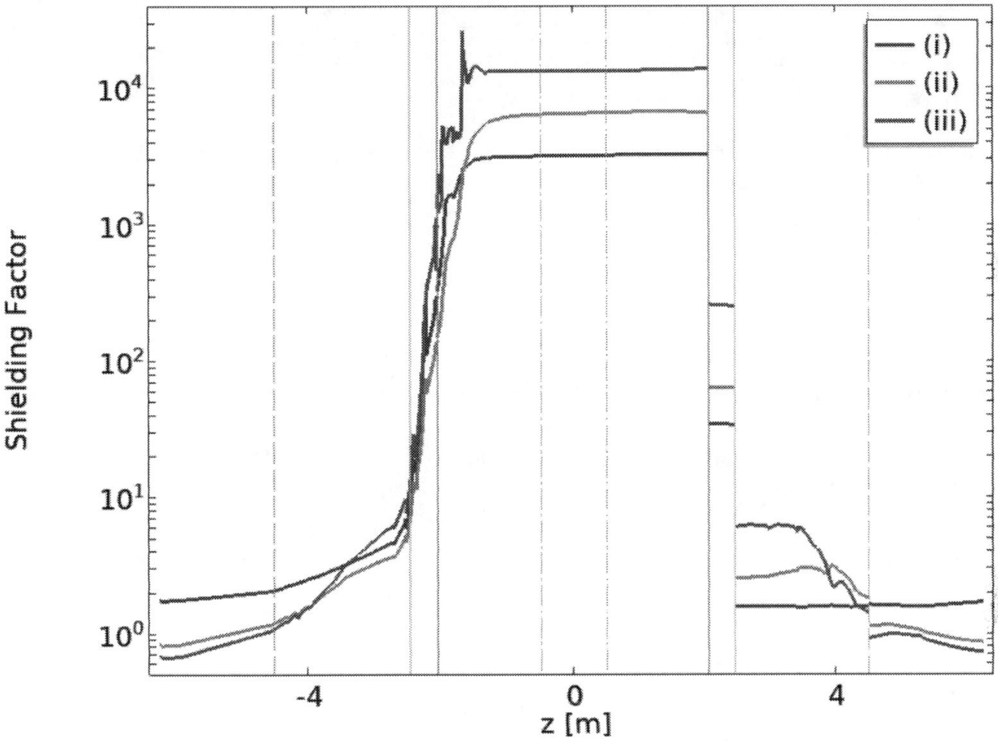

Figure 7. Shielding factor along the axis of the neutron guide for three cases: *i*) No compensation system, *ii*) Compensation system with the original design electric currents, and *iii*) Optimized compensation system. Dash-dot vertical lines indicate the boundaries of the measurement region, solid vertical lines correspond to the inner and outer walls of the MSE, and dashed vertical lines represent the limits of the compensation system.

4 Conclusions

Magnetically Shielded Enclosures, also referred to as Magnetically Shielded Rooms, are essential components in experiments seeking to improve current limits on the nEDM. In this work, we analyzed the performance of an MSE using finite element method simulations in COMSOL Multiphysics® AC/DC Module, focusing on the quality of the magnetic field environment, specifically field uniformity and gradient reduction, as well as the shielding factor achieved. The study considered the influence of ambient magnetic fields in the MSE's immediate surroundings, which are largely determined by the accompanying compensation system.

Our results highlight the importance of optimizing the compensation system to ensure the required field conditions inside the MSE. While the original design was effective to some extent, the optimization process presented here led to significant improvements in magnetic field gradients, uniformity, and overall shielding effectiveness.

Although the experiment for which this study was originally conducted is no longer being pursued at the SNS, this analysis remains relevant. Similar experiments under development worldwide, including the nEDMSF undertaking, which builds upon the approach of the nEDM@SNS experiment [5], face similar challenges in achieving high-performance magnetic shielding, making these findings valuable for future implementations and improvements in MSE designs.

References

[1] R. Golub and Steve K. Lamoreaux, Neutron electric-dipole moment, ultracold neutrons and polarized 3He. Phys. Rep. **237**, (1994). https://doi.org/10.1016/0370-1573(94)90084-1]

[2] C. Abel *et al.*, Measurement of the Permanent Electric Dipole Moment of the Neutron. Phys. Rev. Lett. **124**, (2020). https://doi.org/10.1103/PhysRevLett.124.081803

[3] M. Ahmed *et al.*, A new cryogenic apparatus to search for the neutron electric dipole moment. JINST **14**, (2019). 10.1088/1748-0221/9/11/P11017

[4] Magnetic Field Calculators. https://www.ngdc.noaa.gov/geomag/calculators/magcalc.shtml

[5] V. Cianciolo, Optimizing Spin Dressing Sensitivity for the nEDMSF Experiment. JINST **20**, (2025). https://dx.doi.org/10.1088/1748-0221/20/01/T01006

EPJ Web of Conferences **333**, 03001 (2025) https://doi.org/10.1051/epjconf/202533303001
XLVI Symposium on Nuclear Physics 2025

Iron beam analysis at LEMA-IFUNAM

A. García-Flores[1,*], *A.* Huerta[1], *D. J.* Marín-Lámbarri[1], *E.* Chávez-Lomelí[1], *G.* Reza[1], *C.* Solís[1], *M.* Rodríguez-Ceja[1], and *G.* Ramírez-Pérez[1]

[1]Instituto de Física, Universidad Nacional Autónoma de México, Apartado Postal 20-364, 01000 Ciudad de México, México.

Abstract. At the "Laboratorio Nacional de Espectrometría de Masas con Aceleradores" (LEMA), Institute of Physics of the National Autonomous University of Mexico (IFUNAM) it is possible to produce highly stable, elemental and molecular ion beams with a wide range of energies, depending on the charge state and current intensities, from nA to few μA. In this work we present the characterization of iron (Fe) beam. The isotopic species of Fe was identify by its magnetic rigidity for a given charge state. Energy and mass were confirmed using the Rutherford Back Scattering (RBS) technique.

1 Introduction

Accelerator Mass Spectrometry (AMS) is a highly sensitive technique to measure relative concentrations of a rare isotope to an abundant (stable) one from 10^{-10} to 10^{-15}. The initial purpose for the system was to perform $^{14}C/^{12}C$ measurements accurately. Currently several radionuclides are analyzed with AMS systems around the world, such as ^{10}Be, ^{26}Al, ^{36}Cl, ^{41}Ca, ^{44}Ti, ^{55}Fe, ^{59}Ni, ^{63}Ni, ^{79}Se, ^{92}Nb, ^{93}Zr among others [1–21]. AMS is important to different research areas like earth sciences, anthropology, archaeology, materials science, nuclear physics and astrophysics [22]. LEMA consists of two mass spectrometers coupled to a 1 MV tandem accelerator [23]. The main use of the facility is ^{14}C dating, but also other isotopes are used: ^{10}Be, ^{26}Al, ^{129}I, ^{239}Pu, additionally the system includes a nuclear physics experimental beam line [24].

2 AMS system at LEMA

The isotope separator at LEMA is composed by three sections (Figure 1), a low energy spectrometer, an accelerator section and a high energy spectrometer as in other facilities [24, 25]. The low energy section includes the Sputtering Negative Ion Cesium Source (SNICS), an Electrostatic Spherical Analyzer (ESA) and an injector magnet. Negative ions are produced in the SNICS through atomic ablation and deflected immediately by an 54° ESA and guided to the 90° injection magnet, here, the ions are separated by their magnetic rigidity. In the second section charge exchange and molecular dissociation occurs [26], the negative charged particles beam is accelerated towards the positively charged terminal collides with the gas stripper, loosing all or part of their electrons and generating a charge state distribution which depends on their speed and gas pressure. This process provides positive ions that are conducted to the high energy spectrometer.

*e-mail: araceligf@fisica.unam.mx

https://doi.org/10.1051/epjconf/202533303001

Figure 1. Photograph of the 1 MV Tandem accelerator at LEMA.

The high energy section has two experimental lines, in the first line are included, among other components: a 90° magnet, a fix and a movable Faraday cups, a 120° ESA and a split anode gas ionization chamber to identify and count the isotopes of interest. The second line is composed by a 90° magnet, a movable Faraday cup, a beam profile monitor (BPM) and a chamber for nuclear physics experiments.

3 Experimental procedure and results

Samples were processed at LEMA and the iron cathodes were produced according to the negative ion cookbook recommendations [27] with elemental high-purity materials, 6 mg of the material were pressed into aluminum cathodes with a 5.2 mm external diameter, pre-drilled 1.3 mm diameter by 9 mm deep. A thin gold (Au) layer was deposited on a thick carbon (C) foil with 5 x 10^{17} atoms/cm^2 thickness and area of 1.78 x 1.78 cm^2, mounted in a target holder located into the reaction chamber as shown in Figure 2. A Passivated Implanted Planar Silicon (PIPS) detector was placed at 150° backward angles.

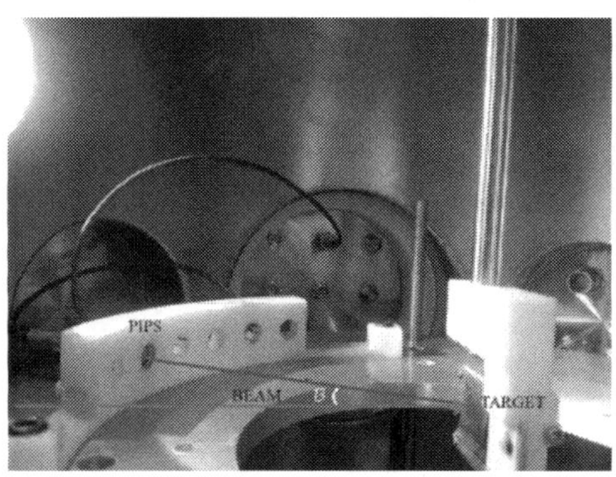

Figure 2. Experimental set-up inside the chamber, where a PIPS detector was placed at 150° backward angles with an exit angle B respect to the beam.

Iron cathodes were placed at the carousel in order to extract the negative iron ions from the SNICS and the beam was focused and filtered in the low energy section. Figure 3 shows a plot of the beam intensity measured by the Faraday cup at the exit of the injector magnet as a function of the mass.

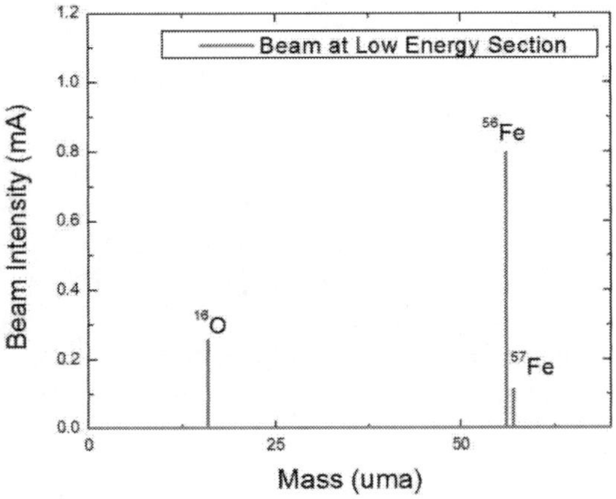

Figure 3. Mass spectrum for a high-purity iron cathode showing the stable ^{56,57}Fe isotopes. Evidently the highest intensity peak corresponds to ^{56}Fe, the most abundant isotope in iron.

After the injector magnet the beam was accelerated in the tandem, and elemental iron positively charged ions were obtained. Some parameters like the magnetic field and the beam current (accuracy better than 1%) corresponding to ^{56}Fe charge states were measured with a gaussmeter and a Faraday cup respectively. The number of particles and the energy (considering the ions initially receive an energy of 35 keV in the low energy section of the AMS system) for each ^{56}Fe charge state were calculated. The resulting values are presented in Table 1.

Table 1. Experimental parameters for the different charge states of ^{56}Fe. The terminal voltage of 1 MV and the gas stripper pressure of 1.8 x 10^{-2} mbar.

Beam	Energy (MeV)	Current (nA)	Intensity (particles/s)	B (G)
$^{56}Fe^{2+}$	3.035	98.0	3.06 x 10^{11}	1704.33 ± 0.17
$^{56}Fe^{3+}$	4.035	35.0	7.28 x 10^{10}	1281.79 ± 0.13
$^{56}Fe^{4+}$	5.035	27.2	4.24 x 10^{10}	1066.22 ± 0.10
$^{56}Fe^{5+}$	6.035	22.5	2.81 x 10^{10}	925.01 ± 0.09
$^{56}Fe^{6+}$	7.035	17.5	1.82 x 10^{10}	826.39 ± 0.08
$^{56}Fe^{7+}$	8.035	12.8	1.14 x 10^{10}	752.01 ± 0.07
$^{56}Fe^{8+}$	9.035	5.3	4.13 x 10^{09}	697.58 ± 0.06

Additionally predictions for the iron charge states distribution using the Schiwietz model were performed, this model improved a formula for the mean equilibrium charge state and introduces a reduced velocity that depends on the projectile and target nuclear charge, also includes recent experimental data for incident ions in solid and gas targets [28], giving an adequate framework for this analysis.

In Figure 4 the ^{56}Fe experimental and theoretical charge state fractions as function of the beam energy and passing through an argon (Ar) target stripper gas are shown. From here the highest probability at 1 MV corresponds to the 2+, followed by the 1+ (which was not measured in this study) charge state and others with lower intensity, the experimental values show an enhancement respect to the theoretical predictions, possibly associated with the data acquisition method and the employed model.

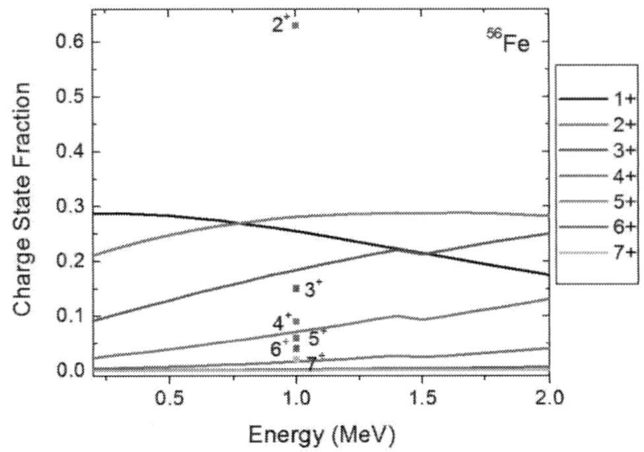

Figure 4. Model predictions (lines) and experimental (squares) values for the charge state fractions in ^{56}Fe as a function of beam energy. In the energy range between 0.2 to 2.0 MeV where 2+ is the most probable charge state at 1 MV.

All iron beams were scattered on the Au on C target. The Simulation Nuclear Reaction Analysis (SIMNRA [29]) was implemented in order to support the experiment, knowing the target was produced through gold deposition on a carbon surface and taking the shape of any peak as a reference, the target was constructed with several layers of Au and C with different concentrations to reproduce its corresponding distribution as shown in Figure 5.

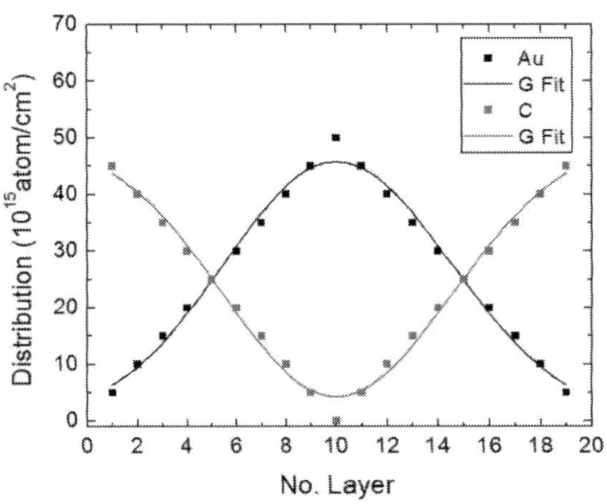

Figure 5. Target simulation with SMINRA with a Au and C layers distribution. The R^2 for the Gaussian Fit (G Fit) was 0.98. The mean surface density for Au corresponds to 16.35x 10^4 ng/cm^2 or 163.54 mg/cm^2 and mean thickness of 86.68 nm or too 5 x 10^{17} atoms/cm^2.

The iron beams on the Au on C target were simulated in SIMNRA with the same set-up configuration and target, varying only the respective beam energy. Experimental spectra were perfectly adjusted with SIMNRA. In Figure 6 the measured and simulated spectra corresponding to $^{56}Fe^{3+}$, $^{56}Fe^{4+}$, $^{56}Fe^{5+}$ and $^{56}Fe^{6+}$ on ^{197}Au are shown. Back scattered particles were recorded with a PIPS detector at scattering angle of 150°.

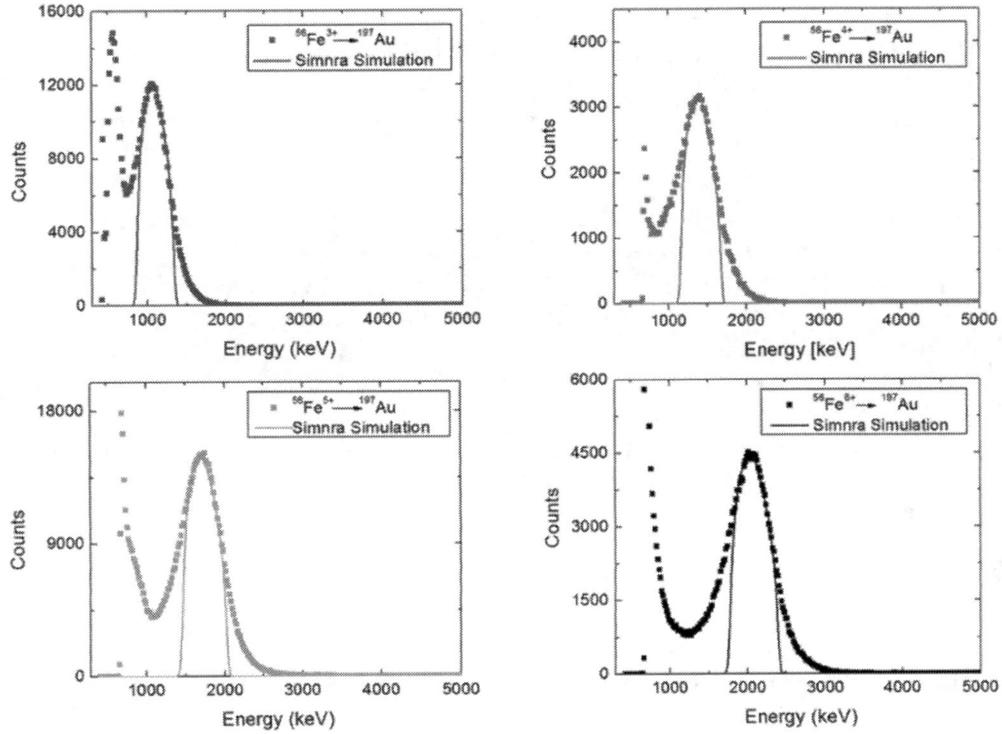

Figure 6. RBS spectra for ^{56}Fe beam on a ^{197}Au target compared to the SIMNRA simulation. In blue, the $^{56}Fe^{3+}$ at 4 MeV, magenta, $^{56}Fe^{4+}$ at 5 MeV, orange, $^{56}Fe^{5+}$ at 6 MeV and black, $^{56}Fe^{6+}$ at 7 MeV.

4 Conclusions

To summarize, we reported the characterization of 2+ to 8+ charge states of ^{56}Fe at the nuclear physics line at LEMA. Theoretical predictions shown that 1+ and 2+ are the most probable charge states at 1 MV. Only the 2+ state was measured experimentally, confirming this forecast. SIMNRA simulations shown the target has not an homogeneous gold distribution. Additionally, RBS measurements shown a good agreement with SIMNRA simulations.

Acknowledgments

We thank to Sergio Martínez González and Carlos Valencia Morales for the technical support. This work has been partially funded by CONAHCYT CBF2023-2024-862, 2023-LNC-58 and UNAM PIIF-5-24, PAPIIT TA10052, IG102023 and IN112023.

References

[1] Kutschera, W. Applications of accelerator mass spectrometry. International Journal of Mass Spectrometry **349-350**, 203-218 (2013). https://doi.org/10.1016/j.ijms.2013.05.023

[2] Wallner, A. et. al. Nuclear astrophysics and AMS- Probing nucleosynthesis in the lab, Nucl. Instrum. Methods Phys. Res., Sect. B **268**, (7-8) 1277-1282 (2010). https://doi.org/10.1016/j.nimb.2009.10.152.

[3] Marín-Lámbarri, D. J. et al. Measurement of the thermal neutron capture cross section by ^{9}Be using the neutron flux from a nuclear research reactor and the AMS technique Phys. Rev. C **102** 044601 (2020).10.1103/PhysRevC.102.044601.

[4] Limata B. et. al. A new study of low-energy (p,γ) resonances on magnesium isotopes, Phys. Rev. C- Nucl. Phys. **82** (1) 1-16 (2010).https://doi.org/10.1103/PhysRevC.82.015801,arXiv:1006.5281..

[5] Zhao Q. et. al. Measurement of the ^{27}Al(n,2n)^{26}Al cross section using accelerator mass spectrometry, Chinese Phys. Lett. **15** (1) 8–9 (1998).

[6] Anderson, T. et. al. Re-measurement of the ^{33}S(α,p)^{36}Cl cross section for early solar system nuclide enrichment, Phys. Rev. C **96** (1) 1-5 (2017). https://doi.org/10.1103/PhysRevC.96.015803.

[7] Arazi A. et. al., Measurement of the ^{25}Mg(p,γ)^{26}Al reaction at stellar energies, New Astron. Rev. **46**, 525-528 (2002).

[8] Martschini M. et. al. Recent advances in AMS of ^{36}Cl with a 3-MV tandem, in: Nucl. Instrum. Methods Phys. Res. Sect. B. **269**, Elsevier B.V. 3188-3191 (2011). https://doi.org/10.1016/j.nimb.2011.04.016

[9] Dillmann I. et. al. Determination of the stellar (n,γ) cross section of ^{40}Ca with accelerator mass spectrometry, Phys. Rev. C- Nucl. Phys. **79**, (6) 1-8 (2009). https://doi.org/10.1103/PhysRevC.79.065805

[10] Hui S.K. et. al.44Ti Atom counting for nuclear astrophysics, Nucl. Instrum. Methods Phys. Res., Sect. B **172**, 642-646 (2000). https://doi.org/10.1016/S0168-583X(00)00101-4.

[11] Paul M. et. al. Counting ^{44}Ti nuclei from the ^{40}Ca(α,γ)^{44}Ti reaction, Nuclear Phys. A **718**, 239c– 242c (2003). https://doi.org/10.1016/S0375-9474(03)00720-6.

[12] Nassar H. et. al. ^{40}Ca(α,γ)^{44}Ti Reaction in the energy regime of supernova nucleosynthesis, Phys. Rev. Lett.**96**,(4) 1-4 (2006). https://doi.org/10.1103/PhysRevLett.96.041102

[13] Coquard L. et. al. Determination of the stellar (n,γ) cross section of ^{54}Fe with accelerator mass spectrometry, in: Int. Symp. Nucl. Astrophys.- Nucl. Cosm. 1-5 (2006). https://doi.org/10.22323/1.028.0274.

[14] Wallner A. et. al. Precise measurement of the thermal and stellar ^{54}Fe(n,γ)^{55}Fe cross sections via accelerator mass spectrometry, Phys. Rev. C **96** 025808 (2017). https://doi.org/10.1103/PhysRevC.96.025808.

[15] Slavkovská Z. et. al. Investigation of ^{54}Fe(n,γ)^{55}Fe and ^{35}Cl(n,γ)^{36}Cl reaction cross sections at keV energies by Accelerator Mass Spectrometry, EPJ Web Conf. **232** 02005 (2020). https://doi.org/10.1051/epjconf/202023202005.

[16] Rugel G. et. al. Measurement of (n,γ) reaction cross sections at stellar energies for ^{58}Ni and ^{78}Se, Nucl. Instrum. Methods Phys. Res. B **259** 683-687 (2007). https://doi.org/10.1016/j.nimb.2007.01.206.

[17] Ludwig P. et. al. Measurement of the stellar ^{58}Ni(n,γ)^{59}Ni cross section with AMS, Phys. Rev. C **95** (3) 035803 (2017). https://doi.org/10.1103/PhysRevC.95.035803.

[18] Dillmann I. et. al. Solving the stellar ^{62}Ni problem with AMS, Nucl. Instrum. Methods Phys. Res., Sect. B **268** 1283-1286 (2010). https://doi.org/10.1016/j.nimb.2009.10.153.

[19] Nassar, H., et. al., Stellar (n,γ) cross section of ^{62}Ni. Phys. Rev. Lett.**94**, 9 (2005). https://doi.org/10.1103/PhysRevLett.94.092504.

[20] He G. Z. et. al. Cross-section measurement for the ^{93}Nb(n,2n)^{92g}Nb reaction at the neutron energy of 14.6 MeV by the AMS method, Chinese Phys. C **42** (7) 1-4 (2018). https://doi.org/10.1088/16741137/42/7/074002.

[21] Pavetich S. et. al. ^{93}Zr Developments at the Heavy Ion Accelerator Facility at ANU, Nucl. Instrum. Methods Phys. Res., Sect. B **438** (3) 77-83 (2019). https://doi.org/10.1016/j.nimb.2018.07.019

[22] Chávez E., et. al. Accelerator mass spectrometry, an ultrasensitive tool to measure cross sections for stellar nucleosynthesis. Nucl. Inst. and Met. in Phys. Res. B **526**, 77-82 (2022). https://doi.org/10.1016/j.nimb.2022.06.016

[23] Solís C. et. al. A new AMS facility in Mexico. Nucl. Inst. and Met. in Phys. Res. B **331**, 233 (2014). https://doi.org/10.1016/j.nimb.2014.02.015.

[24] Reza G. et. al. Characterization of the new hybrid low-energy accelerator facility in Mexico. Eur. Phys. J. Plus **134:590** (2019). https://doi.org/10.1140/epjp/i2019-12950-1.

[25] Kieser, W.E. Accelerator mass spectrometry: an analytical tool with applications for a sustainable society. EPJ Techniques and Instrumentation **10**, 7 (2023). https://doi.org/10.1140/epjti/s40485-023-00088-3

[26] Synal H. A. Accelerator Mass Spectrometry: Ultra-sensitive Detection Technique of Long-lived Radionuclides. Chimia **76**, 45–51 (2022). doi:10.2533/chimia.2021.45

[27] Middleton, R. A negative Ion Cookbook, Department of Physics, University of Pennsylvania (1989). http://www.pelletron.com/cookbook.pdf

[28] Schiwietz, G. and Grande, P.L. Improved charge-state formulas. Nucl. Inst. and Met. in Phys. Res. B 175-177 (2001) 125-131. https://doi.org/10.1016/S0168-583X(00)00583-8

[29] Mayer, M. SIMNRA User's Guide, Technical Report IPP 9/133, MPI für Plasmaphysik, Garching, Germany, 1997. http://home.rzg.mpg.de/ mam/index.html

EPJ Web of Conferences **333**, 03002 (2025) https://doi.org/10.1051/epjconf/202533303002
XLVI Symposium on Nuclear Physics 2025

Current scientific research on Electrostatic accelerator EG-5 in JINR

Alexander S. Doroshkevich[1,2], *Efrain Rafael* Chavez Lomeli[3], *Edwin* Pedrero González[4], *Zhanna* V. Mezentseva[1], *Boris* L. Oksengendler[5], *Tatyana* Yu. Zelenyak[1], *Alisa* A. Tatarinova[1], *Phan Luong* Tuan[1,6], *Vesna* Teofilović[7], *Zoran* Ivanovich[8], *Carmen* Mita[9], *Diana* M. Mardare[10], *Nikoleta* Cornei[10], *Matlab* N. Mirzaev[11,12,13], *Nurbol* O. Appazov[14], *Andriy* K. Kirillov[1], *Ilya* A. Chepurchenko[1], *Anastasiya* I. Kruglyak[1], *Yulia* V. Aleksiayenak[1], *Vitaly* K. Ksenevich[15], *Aleksandr* V. Maletskii[1,16], *Altyn* Zh. Altynbasova[1, 17], *Anthony Carlos* Perez Moreno[1,18], *Rafael* Sh. Isayev[1,19,20], *Mahmoud* Ibrahim[1,21], *Silvia Maria* Fortuné Fábregas[1,22], *Luis Miguel* Ledo Pereda [1,23], *Ilya* O. Simonenko[1,2], *Vladislav* A. Kinev[24], *Alexey* R. Tameev[25]

[1]Joint Institute for Nuclear Research, Dubna, 141980, Russia
[2]Dubna State University, Dubna, 141980, Russia
[3]Física Nuclear y Aplicaciones de la Radiación Instituto de Física, UNAM, Mexico
[4]Institute of Materials Science and Technology (IMRE) of the University of Havana, Havana, Cuba
[5]Institute of Materials Science of the Academy of Sciences of the Republic of Uzbekistan, Tashkent, Uzbekistan
[6]Irradiation Center, Vietnam Atomic Energy Institute, Hanoi, Vietnam
[7]Faculty of Technology, University of Novi Sad, Novi Sad 21000, Serbia
[8]Laboratory of Physics, INN Vinča, University of Belgrade, Belgrade 11351, Serbia
[9]Faculty of Chemistry, “Alexandru Ioan Cuza” University of Iasi, Iasi 700506, Romania
[10]Faculty of Physics, “Alexandru Ioan Cuza” University of Iasi, Iasi 700506, Romania
[11]Institute of Radiation Problems, Ministry of Science and Education Republic of Azerbaijan, Baku, AZ1143, Azerbaijan
[12]Khazar University, Baku, AZ1096, Azerbaijan
[13]Azerbaijan State Oil and Industry University, Scientific-Research Institute Geotecnological Problems of Oil, Gas and Chemistry, AZ1010 Baku, Azerbaijan
[14]Laboratory of Engineering, Korkyt Ata Kyzylorda University, Kyzylorda 120014, Kazakhstan
[15]State Scientific and Production Association "Scientific-Practical Center of the National Academy of Sciences of Belarus", Minsk, Republic of Belarus
[16]Galkin Donetsk Institute for Physics and Engineering, 83114 Donetsk, Russia
[17]Karaganda Industrial University, Temirtau, 101400, Kazakhstan
[18]Voronezh State University, Voronezh, 394018, Russia
[19]National Research Nuclear University “MEPhI”, Moscow, 115409, Russia
[20]IDDA, Department of Nuclear Research, Baku, AZ1069, Azerbaijan
[21]Tomsk Polytechnic University, Tomsk, 634050, Russia
[22]Photovoltaic Research Laboratory, Institute of Science and Technology of Materials (IMRE), University of Havana, Cuba
[23]Center for Technological Applications and Nuclear Development (CEADEN), Havana, Cuba
[24]JSC "Research and Production Enterprise "Integral", Dubna, Russia
[25]Frumkin Institute of Physical Chemistry and Electrochemistry Russian Academy of Sciences, Moscow, 119071, Russia

Abstract. The electrostatic accelerator (ESA) EG-5 has been operating stationary in the Nuclear Physics Department of the Nuclear Physics Department of JINR (Dubna) since 1965. Along with an experimental nuclear reactor and a pulsed accelerator IREN, ESA EG-5 occupies its own unique niche as part of a complex of nuclear physics installations. The beams of high-energy particles obtained using EG-5 have the highest energy stability (± 15 keV per 2 MeV), due to which it is possible to conduct unique studies of the elemental composition of solids, including depth profiling, conducting studies of fast neutron nuclear reactions, etc. ESA EG-5 is a universal research device that allows conducting both studies of the elemental composition and physical, chemical and biological modification of objects of inanimate and living nature, respectively. EG-5 electrostatic accelerator at FLNP JINR, used to produce intense fluxes of fast particles (H^+, He^+, D^+) and neutrons; for elemental analysis of surface layers of various objects using beams of α-particles, using non-destructive techniques RBS, ERD and PIXE; for implantation of ions into the surface layers of various materials; to study the radiation resistance of materials. Unique opportunities will appear after the implementation of a microbeam spectrometer at the EG-5 accelerator in period since 2025.

1 Electrostatic accelerator EG-5

Electrostatic accelerator (ESA) EG-5 - a long-liver laboratory of neutron physics, working properly since 1965. It is a reliable tool for solving a wide range of problems in the field of nuclear physics, solid state physics, radiation technologies. The ion beam of ESA at a relatively high current (up to 100 μA) has a very high energy stability, due to which the EG-5 plant can operate both in the mode of a neutron generator and an ion source for ion beam analysis. Today, with the help of EG-5, studies of nuclear reactions of neutrons with a solid body, accompanied by the departure of charged particles, studies of elementary depth profiles of multilayer semiconductor architectures, as well as irradiation (radiation modification) of materials are carried out.

Since 2023, the electrostatic accelerator has been the primary installation of the FNLP. It is a hardware complex that includes a high-voltage (up to 5MeV), a high-pressure gas cylinder system, a vacuum, electromagnetic ion beam control system, an electronic control system and automation of technological processes. The equipment of the complex is located in the building of the accelerator "tower" to 42 and the premises of two adjacent experimental halls (Fig. 1).

Fig. 1. EG-5 accelerator complex. 1 – Chamber for irradiation of silicon wafers (JSC Mikron); 2 – Chamber for irradiation with singly charged ions; 3 – Installation for studying reactions with the release of charged particles; 4 - Chamber of ion beam spectrometers; 5 – Microbeam spectrometer (within a few years); 6 – Installation for research using the channeling method [1].

The accelerated ion beam is distributed through ion pipelines to one of 6 channels (three for each experimental hall) equipped with terminal devices. Terminal devices are separate nuclear-physical installations, each of which is quite unique both within JINR and JINR Member States and Associate Members.

Naturally, among the accelerator units of the Joint Institute, EG-5 occupies its own unique niche.

Significant advantages of Electrostatic accelerator EG-5:
- high energy stability of ion beam;
- high intensity of ion beam;
- accelerated particles (H^+, He^+, D^+);
- accelerated voltage (from 1,1 MV to 3,5MV).

Areas of use of Electrostatic accelerator EG-5:
- Nuclear reactions with fast quasimonoenergetic neutrons;
- Ion Beam Spectrometry (Multilayer structures, isotope determination, elemental depth profiling);
- Radiation technologies (Science, technology, medicine, etc.).

Ion beam parameters of EG-5:
- Range of ion beam currents 0,01 - 30 μA (100 – 150 μA*);
- Ion beam energy range 1,1 – 3,5 MeV (4,1 MeV*);
- Energy resolution (H^+, He^+) - 15 keV;
- Charged particles flow (H^+, He^+) – 10^{12}–10^{13} part /s cm^{-2};
- Neutrons flow $5 \cdot 10^7$ pat/s cm^2;
- Neutrons energy 20 – 800 keV; 3,5 - 5,1MeV ± 0,1 MeV.

*- will be after modernization

The accelerator complex is not limited to ion beam spectrometers and irradiation chambers (Fig. 1). There is also a chemical laboratory synthesizing organic-inorganic perovskite structures (for solar energy harvesting and X-ray detection) and semiconductor CuO layers (tailored for nanostructured material-based solar cells), alongside an engineering center dedicated to developing functional architectures for nanoelectronics, nanosensors and renewable energy applications.

Appointment of the EG-5

Currently, particle energies up to 4-5 MeV are no longer of great importance for fundamental nuclear physics, but a range of various applications where EG-5 beams are needed, such as, in particular, bionanotechnology and interdisciplinary scientific fields. These studies are in demand, as evidenced by requests for results from international rating journals.

The "center of mass" of the spectrum of actual scientific problems for EG-5 has shifted in the field of applied research since its inception (1965). At the moment, several scientific areas have emerged in the sector, covering a wide range of the most pressing tasks of our time: from the development of promising electronics devices (homogeneous nanoelectronics) and the production of drought-resistant rice varieties to the study of astrophysical processes of stellar nucleosynthesis [2, 3, 4].

EG-5 unique opportunities

The accelerator system based on ESA EG-5 is multifunctional. In appropriate modes, it can be used as a source of ionizing radiation, as a tool for carrying out physical and chemical modification of materials, and, of course, as a spectrometric research instrument. For example, using fast neutrons with a narrow energy dispersion obtained by the (d-d) reaction on the accelerator EG-5 relatively high doses of radiation can be given to the test materials without inducing radioactivity in them. By implanting ions of inert helium gas, it is possible to significantly change in localized volumes of materials (near-surface layers) the field of mechanical stresses, than to induce structural-phase transformations or create the so-called "helium porosity." In the latter case, there is also a change in microstructure and atomic density. Helium porosity, like other radiation-induced defects, is a serious problem of radiation technologies and has always been a relevant subject of research [5, 6]. It should be noted that EG-5 is currently the only accelerator in JINR that allows such studies on helium in the energy range of 1-3.5 MeV. Implantation of hydrogen ions allows in strictly localized regions of the material to carry out partial chemical reduction of metal oxides, than to change the stoichiometric ratio of elements, i.e. to carry out chemical modification of materials. Implantation of ions into the crystal lattice of materials also allows changing their electronic structure and electrical properties, which is actively used in the electronic industry [7]. A significant advantage of accelerated implantation technologies is the accurate spatial localization of the implanted ions, and the presence of intermediate layers of material in the path of the ion beam is not a significant obstacle for this [8, 9].

Irradiation of biological objects with fast neutrons allows causing useful mutations for new varieties of agricultural products [4, 66, 67, 68]. Studies of neutron-induced reactions with the escape of charged particles provide valuable information about the mechanism of nuclear reactions and the structure of the atomic nucleus, the processes of stellar nucleosynthesis, etc. Recent significant results, indicating the need to clarify the existing ideas about the main neutron source in astrophysical processes - the $^{22}Ne(\alpha,n)^{25}Mg$ reaction, were obtained at EG-5 under the guidance of Professor Yu.M. Gledenov [1].

Within the last three years with use of EG-5 on fast neutrons sections (n, α) - reactions for ^{144}Sm, ^{66}Zn, ^{10}B, ^{25}Mg, $^{54,56}Fe$, $^{58,60,61}Ni$ elements were measured and at the moment together with Obninsk IPPE are conducted works on The Neutron Data Library BROND on a number of nucleus (^{6}Li, ^{14}N, ^{35}Cl, ^{91}Zr, ^{56}Fe) [10, 11].

Research of nuclear reactions

At the Van de Graaff accelerator of the FLNP JINR, the cross section for the reaction $^{35}Cl(n,\alpha)^{32}P$ was measured at a neutron energy in the range of 3.3-5.3 MeV [12].

Chlorine, widely distributed in nature, is one of the chemical elements necessary for man, and the content of the ^{35}Cl isotope is 75.77%. An experimental study of this reaction could complement the existing library of nuclear data and can be used to test the corresponding models of nuclear reactions, as well as to refine the parameters of the optical potential used in calculations of various astrophysical scenarios. The cross sections for the $^{35}Cl(n,\alpha)^{32}P$ reaction are useful for applications in nuclear technology. For example, they were used, to estimate the fast neutron flux during the JCO accident at the Tokaimura nuclear facility in 1999. The beta-radioactive product ^{32}P can be used to study the absorption of phosphate fertilizers by plants by radionuclide tracking. In addition, chloride is a component of the coolant of a molten-salt reactor, which is one of the six most promising GEN-IV reactors.

The physics of scattering of monoenergetic alpha particles by atoms of near-surface layers of matter is the basis of ion beam spectrometry methods. The uniqueness of these methods is due to the ability to study multilayer planar structures without destroying them by layer thinning, as, for example, when using the XPS method. Moreover, the sensitivity limit of ion-beam analysis is at least two orders of magnitude higher than XPS (is of the order of 10^{-3} – 10^{-4} atom %) [13]. It should be noted that ion-beam analysis methods allow for quantitative elemental analysis of light atoms, as well as isotopes. Using the EG-5 installation on helium ion beams, unique non-destructive experimental studies of elemental depth profiles with a depth resolution of about 10 nanometers are carried out. (RBS method) [14]. The high sensitivity of the method (up to 10^{15} atoms/cm^2 [15]) allows, for example, to determine the content of an impurity of heavy elements in an amount of 0.001 at.% [16] or to recognize a substance in the form of a layer up to 1 nm thick. In complex use, non-destructive testing methods (Elastic Recoil Detection (ERD), Nuclear Reaction Analysis (NRA), Particle Induced X-Ray Emission (PIXE)) based on proton and helium ion beams with energies from 1 to 3.1 MeV make it possible to determine with high accuracy the elemental composition and depth distribution in the near-surface layer of all elements of the Mendeleev's Periodic table and their isotopes (including hydrogen).

Characterization of elemental composition of materials

The elemental composition of glass/ITO/TiO_2/$CH_3NH_3PbI_3$-type functional structures for solid-state X-ray detectors in the form of thin films was investigated by XRD, RBS methods. The high spatial homogeneity of the obtained structures and the possibility of their further use for the manufacture of radiation-resistant sensitive elements of a position-sensitive X-ray detector are shown [17].

Subnanoporosity of CaF_2-doped ErF_3 crystal by Rutherford Backscattering data

With use of a method of Rutherford Backscattering Spectrometry (RBS) the porosity of optical crystals of CaF_2 alloyed by ErF_3 (2.0 and 5.5%, respectively), received by the Bridgman method was investigated at the atomic level. The volume fraction and diameter of the pores present were found to range from 18 to 22% and from $2.4 \cdot 10^{18}$ to $3.0 \cdot 10^{18}$ atoms/cm^2, respectively [3].

He^+ ion implantation was applied to modify the structural properties of tin oxide films. On the basis of the X-ray diffraction analysis, it was found that both the initial and ion-implanted films are characterized by a multiphase structure consisting of SnO, SnO_2, and Sn_2O_3 phases. By means of the RBS method it was established that the implantation of He^+ ions with energy E = 2.4 MeV at an irradiation dose of $1.1 \cdot 10^{16}$ cm^{-2} into disordered films of tin oxides does not induce a significant redistribution between the content of Sn and O atoms through the whole films thickness [18].

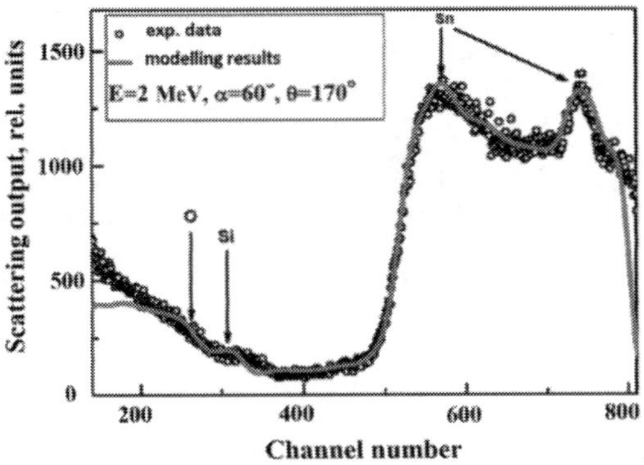

Fig. 2. Energy spectrum of tin oxides film in which He^+ ions with energy E = 2.4 MeV are implanted.

Surface and subsurface defects of dental alloys exposed to sandblasting have been investigated by using RBS and variable energy positron beam studies indicate shallow alumina deposition in material and show that small air pressure of 0.1 MPa is not enough to remove the most metal surface oxides completely in 60 s in all studied dental alloys [19]. It was observed that higher pressure causes the increase in roughness and damaged zone range. The type of defects was determined as vacancies and dislocations. The defect concentration decreases with the depth and depends on alloy's type and applied air pressure.

The irradiation-induced defects and their spatial distribution in pure (99.99% purity) polycrystalline iron samples after exposing of proton implantation with energies of 73 keV and 173 keV and doses of 3×10^{15}, 10^{16} and 3×10^{16} cm^{-2} were investigated by using Elastic Recoil Detection (ERD) and Variable Energy Positron beam (VEP) methods [9]. Analysis of VEP results was performed with VEPFIT code. The strong reduction of positron diffusion lengths in the implanted layers point out the presence of defects with high concentration. The thicknesses of these layers are in good agreement with hydrogen range evaluated by ERD. However, the presence of defects with lower concentration behind implanted zone was also detected when the free fitting was performed. The long-range effect cannot be excluded.

Using the technical capabilities of the EG-5 the fabrication by Swift Heavy-Ion (SHI) bombardment of nanometer-sized reduced graphene oxide (rGO) spots which can be considered graphene quantum dots (QDs) embedded in a non-conducting matrix was found. It were shown that both the number density and the diameter of the rGO spots can be tailored by a suitable choice of irradiation parameters (i.e., ion type, fluence, and energy). It was found that the degree of graphene oxide defunctionalisation by SHIs with different energies scaled well with the deposited electronic energy density. It was shown that the resistance of the samples decreased nonlinearly with increasing ion dose and, at fluences above 10^{13} ions/cm^2, was orders of magnitude lower than the initial value. It was found that an increase in the electronic stopping power of the ion resulted in suppression of the structural ordering at low fluences and in increased amorphization efficiency and formation of *sp*-hybridized carbon chains of both polyynes and polycumulenes at high influences. A hypothesis suggesting that the *sp*-C chains are bridges joining opposite sides of nanoholes created inside the track core and thus assuming the formation of a coupled QD-antidot system is presented. These phenomena were found to be absent in comparative experiments with 200 keV Xe ion irradiation, i.e., in the nuclear stopping regime [20].

Investigation the changes of mixing degree for implanted TiO_2/SiO_2 with changing of energy and mass of incident ion

The investigation of depth distribution of elements in the multilayer structures of $TiO_2/SiO_2/Si$ before and after ion irradiation was provided by using of EG-5 facilities. On the samples which were implanted with Ne^+, Ar^+, Kr^+ or Xe^+ ions with the energy 100, 150, 200 and 250 keV the transition layers existing between the TiO_2 and SiO_2 layers was found by using the Rutherford Backscattering Spectrometry method. Formation of these layers derived from the ion beam mixing that was occurred at TiO_2/SiO_2 interface after irradiation process. The depth profiles show that thickness of the transition layers increased with the growing energy and atomic mass of the implanted ions [5, 21]. The variation of the transition layer thicknesses of TiO_2/SiO_2 before and after the ion irradiation were investigated using the relative thickness. The values increase with increasing of ion's energy and mass, this behaviour corresponds with increasing of the energy transfer to the recoil atoms in the transition layers (calculated by SRIM software). Based on a comparison, contribution of decreasing FWHM to mixing is not significant, in contrast the displacement of the Ti atoms across the interface is predominant contributing to the broadening of the mixed layers towards the substrate more than that in the opposite direction.

Investigation of changing in chemical composition and optical properties of TiO_2/SiO_2 before and after irradiation

In addition to ion beam spectrometry, the group actively uses complementary methods for studying the physical properties of the surface, in particular, spectroscopic ellipsometry, impedance metering and volt-amperometry. These methods make it possible to characterize the electrical, electronic and optical properties of the studied objects.

The XPS spectra obtained from the original and implanted samples have different shapes and the intensity of the bands changes. It was found that under the influence of impinging ions, the intensity of the Ti and TiO_2 bands decreases, however, that of TiO and Ti_2O_3 increases with the increasing of energy from 100 to 250 keV. Decreasing concentration of TiO_2 after irradiation lead to increasing of deposited energy which is reflected in increasing amount of mixing or broadening of transition layers, this shows a good agreement with RBS method. Ellipsometric measurements of the material before and after ion implantation show a correlation between in the yields of the bands and the angle of incident ion beam. This is associated with the interference processes of light reflected from the examined structure. The refractive index n and the extinction coefficient k have been found to increase after implantation up to 200 keV then to decrease at 250 keV Xe ion. Furthermore, the increasing of the thickness of transition layers between the TiO_2 and SiO_2 materials that has been observed in the ellipsometric measurements are in good agreement with the RBS results [22].

The study of the effect of 250 keV Xe^+ implanted to GaAs samples with the ion fluence from 1×10^{12} to 3×10^{16} ions/cm^2 was made [8]. The spectroscopic ellipsometry approach reveals that the pseudo-dielectric function of implanted GaAs samples and the RBS exposes the depth-profile of As, Ga, and Xe in the implanted samples. The Nuclear reaction (NR) points out the existence of an oxygen-enriched layer on the surface of implanted GaAs samples.

An RBS/NR study was verified the presence of an relatively dense oxygen layers on the surface of the Xe^+ ion implanted GaAs samples with the fluence 3×10^{16} cm^{-2} in comparison with original sample. The changing in chemical composition of Ga and As atoms in the near surface layer and the presence of Xe in the samples reflects in the form of the changing of the channel number on the spectrums from 700 to 850 (Fig. 3, a).

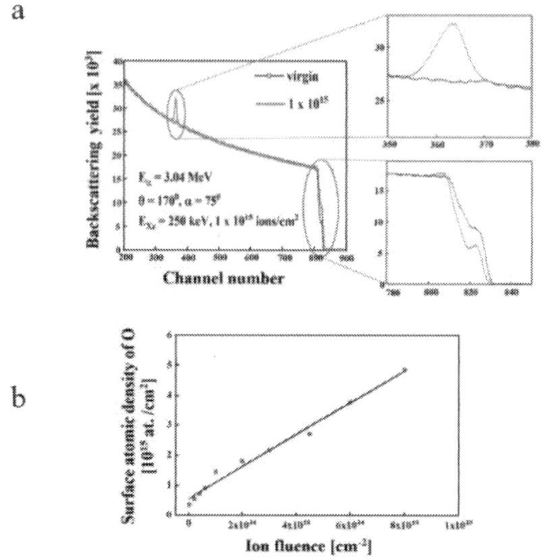

Fig. 3. RBS/NR spectra of samples before and after implanted.

This indicates a difference in diffusion process of Ga and As atoms during the ion irradiation with different fluence. The embedding of the implanted element (Xe element) into GaAs samples causes the change of atomic density. These changes are related to the increasing of the vacancies formed in the near surface layers of implanted GaAs. The number of vacancies produced during irradiation depend on the ion fluence. The thicknesses of the native oxide layers were determined based on RBS/NR spectra due to performing measurements at different energies of incident α particles. Was found that the thickness of the native oxide layer is a function is a linear function of the fluence in the range from 2×10^{13} up to 8×10^{14} ions/cm^2.

Engineering Oxide and Semiconductor Structures with Ion Beams

On semiconductor materials, together with the Department of Semiconductor Electronics and Semiconductor Physics, NUST MISIS, Moscow, a number of successful experiments were carried out with the irradiation of semiconductor crystals [6, 23, 24].

Reverse current recovery time is an important characteristic of diodes and transistors, determining their operating frequency and application. Defects in the semiconductor structure can significantly delay the recovery of the reverse current, reducing the speed of the device and leading to additional heating. Papers [6, 23, 24] present the results of determining the recovery time of a small reverse current in Schottky diodes based on α-Ga_2O_3 (Sn), β-Ga_2O_3, γ/β-Ga_2O_3 and NiO/Ga_2O_3 films. Measurements were made by applying current pulses of different polarity to the diode and taking a reverse current relaxation curve. It has been shown that in diodes based on the above films on their own substrate, the characteristic time of reverse restoration of small currents can reach 20 μs and is limited mainly by the characteristic decay time for the RC chain formed by the structure.

Irradiation of structures such as α-Ga_2O_3 (Sn), β-Ga_2O_3, γ/β-Ga_2O_3 and NiO/Ga_2O_3 with protons with an energy of 1.1 MeV leads to the appearance of a large number of defects inside the structure that can "work" as deep centers of carrier capture. In this case, thermal ejection of carriers from such centers can delay the relaxation of the reverse current recovery, slowing down the switching process. In the α-Ga_2O_3 sample after irradiation, the characteristic switching time was 6 μs, which is twice the time that would be expected based on the parameters of the equivalent RC chain.

The delay in the relaxation of the reverse current is determined not only by the discharge time of the RC chain, but also by the large number of deep centers that can be introduced by irradiation or occur during the growth or operation of the device. Using admittance spectroscopy and deep-level relaxation spectroscopy, data were obtained on the nature of the capture centers that slow down the process of switching the diode at room temperature: these are probably interstitial oxygen atoms. The obtained results can be used to improve crystal growth technology for producing Schottky diodes with high boundary frequency.

1.1 Powder Nanotechnologies

Functional transition for homogeneous electronics technologies

For the first time, a rectifying contact in the form of chemically homogeneous hydrated nanoparticles of the ZrO_2 –Y_2O_3 system, which is based on the dimensional effects of distortion of the zone structure by the surface of low-dimensional objects has been practically implemented and investigated. The use of new physical principles to obtain a rectifying contact on the specified model object opens up prospects for solving the problem of diffusion instability of chemically inhomogeneous heterostructures, which significantly limits the applicability of classical semiconductor devices under conditions of elevated temperatures and ionizing radiation. The prospects of creating homogeneous electronics devices that allow operation in harsh physical conditions, with optical transparency, mechanical strength, biocompatibility (in the case of using YSZ systems) and the ability to scale into a submicroscopic size range are shown [2]. Prospects for creation of homogeneous electronics devices of considerable interest for critical technologies.

In the framework of the Tamm theory of surface states and the generalized Seitz–Madelung model, the electronic structure of the near-surface states of nanoscale ion crystals was modeled with regard to ionic crystals, taking into account the Coulomb range and dispersion of electronic properties depending on the curvature of their surface (Fig. 4,b) [25]. The possibility of realizing the varizonicity (graded band-gap) of the electronic structure in chemically homogeneous ionic crystals, predicted earlier in [26, 27] was shown. The possibility of using the obtained structure to convert the energy of moisture adsorption into an electrical form was shown.

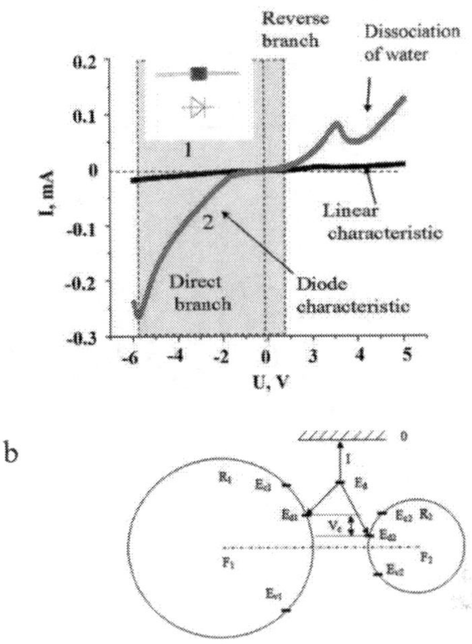

b

a $V_c = -1/2(E_{d1}-E_{d2}) \neq 0$

$$E_d = \frac{1}{2}E_g\left[1-2me^4/E_g\xi^2\hbar^2\right]^{1/2}$$

Fig. 4. The VAC of the nanoparticle contacts with the same (7.5 nm, curve 1) and different (7.5 and 9 nm, curve 2) nanoparticle diameters **(a)** and the schematic interpretation of the rectifying contact effect **(b)**.

1.2 Nanopowder metal-oxide ionistor

As part of studies of the processes of charge and material exchange of nanoscale systems with the external environment, a new dimensional effect of accumulation by the system of compacted zirconia nanoparticles (ZrO_2-3% mol Y_2O_3) of an electric charge with a density of up to 270 μF/g during exposure in an electric field (5000 V/m) under normal physical conditions was established [28]. Based on a qualitative complex analysis, it was shown that the object of localization of different charge carriers is the surface of nanoparticles. In the framework of the theory of dispersed systems, zone theory and the theory of contact phenomena in semiconductors, the proposed mechanism of effect is considered [29, 30]. It was concluded that it is caused by the phenomenon of localization of electron-type charge nanoparticles L^1 in the near-surface zone of the material (Fig. 5) in contact with the adsorption ion atmosphere. The effect is relevant for modern subvolt nanoelectronics, microsystem technology.

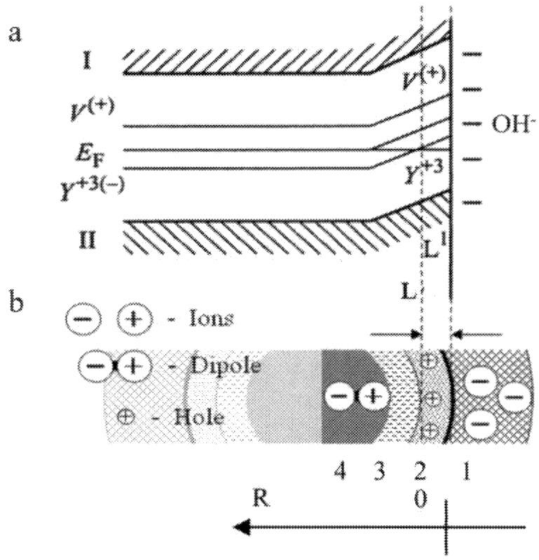

Fig. 5. (a) Qualitative band model of the surface of the ZrO_2-Y_2O_3 system with chemically adsorbed quasi-water molecule (OH^-). The surface is negatively charged.
I – conduction band;
II — valence band.
(b) Spatial distribution of charge in the nanoparticle volume and beyond.
1. Hydration shell.
2. Hole conductivity layer (L^1).
3. Field penetration area (L).
4. Neutral dielectric core of the nanoparticle.
R is the spatial coordinate [29].

In the framework of joint studies of dimensional phenomena, theoretical work to study the effect of radiation on the conformation of DNA molecules [31, 32]; processes of self-organization of nanopowder systems based on ZrO_2 in conditions of high hydrostatic pressure [33]; influence of group IV elements on sintering kinetics of ZrO_2 nanopowders [34], on the development of semiconductor heterostructures based on linearly ordered nanoparticle ensembles [35], heterodyne frequency spectrum converters for bionanoelectronics devices [36], nanosensorics devices [37], organo-inorganic perovskite photovoltaics [38-42] were conducted. Important scientific tasks of clarifying of nuclear data on the reactions of stellar nucleosynthesis and the study of the mechanisms of the influence of cosmic radiation on the evolution of animate organisms of the Earth on the example of rice seeds [43].

2 Radiation materials science

The admixture of helium atoms in materials is a typical product of (n, α) transmutation reactions initiated by neutron irradiation. Controlled introduction of impurity He into the test material using an ion beam is extremely promising as an alternative to long, complex and expensive experiments in a research nuclear reactor. The main problem with ion beam experiments is the significant heterogeneity of the ion doping profile, and therefore there is a certain difficulty in determining a relationship between the observed structural changes and the amount of impurity incorporated. With the use of a special target unit, which makes it possible to solve this problem by obtaining an ion doping profile in the sample that is uniform in the depth of stopping the embedded ions at the LNP accelerator EG-5 uniform doping with He ions (2,5 MeV) was carried out for the first time model samples made of 08Kh13 steel, the principle [44] and the design of a special target unit and software for determining the accumulated ion doping profile in real time of the experiment were tested. The helium-embedded steel samples obtained in the experiment were then used for electron microscopic observation of the development of gas porosity (during annealing) in helium-doped ferritic steel to a given concentration.

Light ion irradiation opens up new possibilities in controlling the properties of semiconductor materials. As part of work with the Arifov Institute of Ion-Plasma and Laser Technologies, studies were carried out on the radiation resistance of a metastable α-Ga_2O_3 that is promising for applications in radiation-resistant devices for opto-, micro- and nanoelectronics. Using the Rutherford Backscattering (RBS) method, voltammetry and impedance spectroscopy, the mechanism of radiation-induced processes in semi-insulating films of α-Ga_2O_3 and β-Ga_2O_3 was investigated. It has been shown that the radiation resistance of these materials relative to electrical properties exceeds the radiation resistance of the advanced material GaN. The determining role in the formation of radiation properties of the specified materials of Ga-vacancies, divacancy defects of Ga-O type and processes of their segregation near the surface is shown [45].

In single-crystal silicon samples and *n*-type films doped with P and Pt and Sn obtained by vapor deposition (by HVPE method), the effect of proton irradiation on electrical properties depending on the donor concentration was studied [46]. The effect of proton irradiation on the concentration of optically active oxygen in silicon samples was studied, and the types of radiation-induced defects were determined.

Due to the severe consequences of accidents with loss of coolant (abbreviated as LOCA - Loss of Coolant Accident), research is underway in the world to create emergency-resistant fuels (ATF). One of the directions is the preservation of zirconium-based alloys as fuel cladding materials by creating corrosion-resistant coatings on their surface in contact with the coolant. One of the materials studied as a coating is chromium, the oxide of which is Cr_2O_3 an effective oxygen barrier.

As part of joint work in the field of radiation materials science with specialized organizations, we have developed a unique technology for obtaining radiation-resistant chemical-inert coatings [47], radiation resistance studies of the corresponding ceramic materials [48, 49], non-stoichiometric TiZrCN coatings were carried out, TiZrNbCN and TiZrSiCN, it has been established that there are no changes in the morphology of coatings after irradiation with protons with an energy of 2.0 MeV at flows up to 10^{17} ion/cm^2, a promising method for protecting the inner surface of fuel element claddings made of EP823-Sh steel has been developed. The mechanism of liquid metal corrosion of chromium steels of EP823-Sh type of ferrite-martensitic class used in reactor technologies has been investigated. Selective dissolution of steel components from the surface when interacting with molten lead under radiation conditions was recorded.

A significant dependence of the process on the oxygen content on the contact surface of the liquid and solid phases has been determined. It is shown that if oxygen concentration is less than stoichiometric value, corrosion products are formation of external oxide layer of magnetite, chromium spinel of non-stoichiometric composition of $Fe(Fe_{1-x},Cr_x)_2O_4$ type. The development of the internal oxidation zone along the grain boundaries has been established [50]. A unique method has been developed to prevent radiation-induced corrosion of the inner surfaces of steel fuel shells made of EP823-Sh steel with mixed nitride fuel, planned for use in the reactor BREST-300-OD by forming a special coating based on chromium [51].

The results of corrosion tests of chromium coating applied to the inner surface of the EP823-Sh steel shell by electrolytic deposition are presented. Electron microscopic studies of chromium coating on steel EP823-Sh did not reveal significant signs of corrosion damage during tests in the environment of fission product simulators (CsI + Te) and in liquid lead at a temperature of 650°C. Analysis of the interaction in the "coating-steel" system showed the prospect of using chromium as a protective coating on the inner surface of the fuel element cladding made of EP823-Sh steel [52].

Alloying with Si or Nb has not been found to affect the grain size of ceramic coatings, but leads to "swelling" of the grains after proton irradiation due to the accumulation of defects in the form of vacancies. All tested TiZrCN, TiZrSiCN and TiZrNbCN coatings showed good radiation resistance.

The effect of proton irradiation with a power of $1\cdot10^{17}$ units/cm^2 and an energy of 2 MeV on the structure and properties of composite ceramics of the composition ZrO_2–SiO_2–Al_2O_3 is shown [53]. It has been determined that at this irradiation dose, the phase composition of the ceramic does not change. Calculations using X-ray diffraction methods have shown that proton irradiation creates compressive stresses (stresses of the 1st kind) ranging from ~ –1 to –2 GPa on the surface of field ceramics, while microstresses (stresses of the 2nd kind) are practically absent. Analysis of SEM images of the ceramic surface after irradiation showed a chaotic arrangement of macropores in the t–ZrO_2 matrix, while pores in zircon particles are located exclusively along the boundaries of inclusions. A decrease in the level of hardness and density in ceramics after proton treatment was noted due to the formation of a large number of pores [54].

The work at the ESA EG-5 is underway to obtain radiation-resistant ceramics. The effect of processing powders with high hydrostatic pressure (HHP) (300, 500, 700 MPa) and stabilized zirconium dioxide (ZrO_2 + 3 mol.% Y_2O_3, (YSZ)) doping on the compaction of metastable nanopowders mixture with γ/θ-Al_2O_3 + n% YSZ (n = 0, 1, 5, 10, 15wt.%) composition during sintering was analyzed. It has been determined that sintering of compacts occurs in two stages: compaction of γ + θ-Al_2O_3 (I-st stage metastable phases), compaction of the stable α-Al_2O_3 (II-nd stage), accompanied by inhibition between them. A connection between the above mentioned inhibition and an increase in the YSZ additive was found. That, also, correlates to the effect of "mutual protection against crystallization" of Al_2O_3 – YSZ powder mixtures. A study of the mechanisms and activation energies of γ/θ-Al_2O_3 + n% YSZ systems' (n = 0, 1, 5, 10, 15 wt.%) sintering depending on the YSZ additive amount and the HHP value showed that a dopant and pressure increase leads to activation energy decrease of the initial sintering stage, and the prevailing sintering mechanism is volumetric diffusion in corundum grains. Also, the work shows that for γ/θ-Al_2O_3 + ≥5wt.% YSZ systems, the optimal compaction pressure is HHP ≥ 500 MPa, which corresponds to the transition of the sintering mechanism from grain-boundary diffusion to volumetric diffusion.

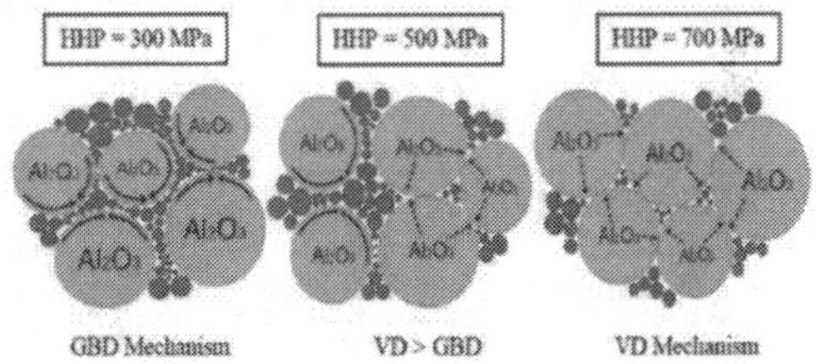

Fig. 6. Schematic representation of the effect of HHP on the sintering mechanism and the distribution of YSZ (at a concentration of ≥5wt.%) in the aluminum oxide matrix at the initial stage of sintering (● - YSZ; ● - Al_2O_3).

The paper [55] presents the results of the study of the structure and physico-mechanical properties of ceramics composites α-Al_2O_3 + n Y_2O_3 (n = 0; 0.5; 1; 1.5; 2; 3; 4; 5 wt.%) based on the polymorphic modifications γ/θ-Al_2O_3 depending on the concentration of the Y_2O_3 doping impurity and annealing temperature of the powder mixtures (800 and 900°C). The effect of mutual protection against crystallization was discovered, which results in mutual inhibition of crystallization processes in Al_2O_3–Y_2O_3 powder systems. The formation of a phase of yttrium-aluminum garnet $Y_3Al_5O_{12}$ (YAG) in ceramics has been established by X-ray diffraction analysis (Fig. 7). The dependence of the mechanical characteristics of the materials under study on the amount and size of the formed phase YAG has been revealed.

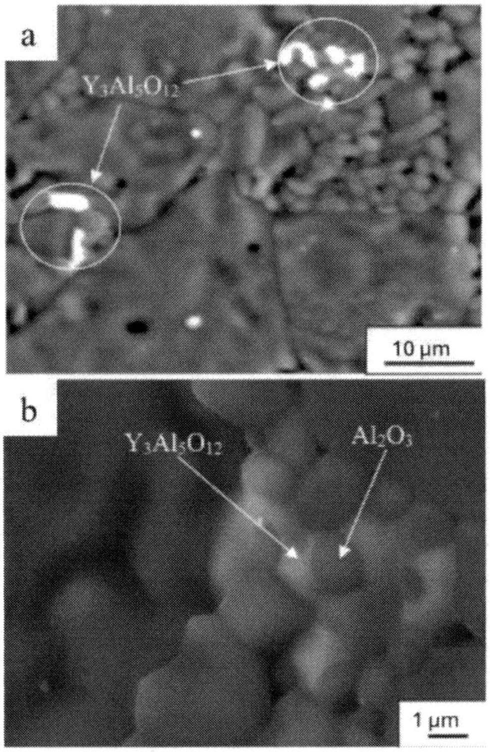

Fig. 7. Surface structure of ceramic samples of the composition Al_2O_3 + 0.5% Y_2O_3, sintered at a temperature of 1550°C from powders annealed at 800°C.

Neutron irradiation of plant seeds is a technique used to induce genetic mutations, often for the purpose of developing new plant varieties with desirable traits such as disease resistance or stress resistance, increased yield, and nutritional content [56-59]. However, successful application in breeding programs requires careful optimization of irradiation doses and a deep understanding of the balance between mutation induction and genomic stability. We conduct neutron irradiation studies on two cultures - rice *(Oryza Sativa)* and triticale *(× triticosecale)* at the EG-5 accelerator. Rice irradiation allowed obtaining mutant forms resistant to salinity and drought factors. Mutants forms of M3 generation with desired properties will be used in further in synthetic rice breeding [60, 61]. Triticale irradiation helped to induct plant variability [62]. These findings collectively highlight the potential of neutron irradiation to contribute significantly to advances in plant genetics and breeding.

As part of the work on the development of new-generation X-ray detectors for medical applications, the radiation resistance of new architectures based on organo-inorganic perovskites such as: PEDOT:PSS | P-TPD (30nm)| PVK (5nm) | G-22 QD (1) и PEDOT:PSS | ---| --- | G-22 QD | C60 | BCP (2), provided by the Frumkin Institute of Physical Chemistry and Electrochemistry of the Russian Academy of Sciences.

The radiation resistance of structures under conditions of continuous irradiation for 60 min was determined by Voltammetry (Potentiostat P-45X, Ellins. Chernogolovka, Russian Federation). The family of obtained V-I curves are shown on Figure 8. It can be seen that during the irradiation time with relatively intense X-ray fluxes of 1.7 x 10^{14} photons / s cm^2, the sample was resistant to radiation exposure.

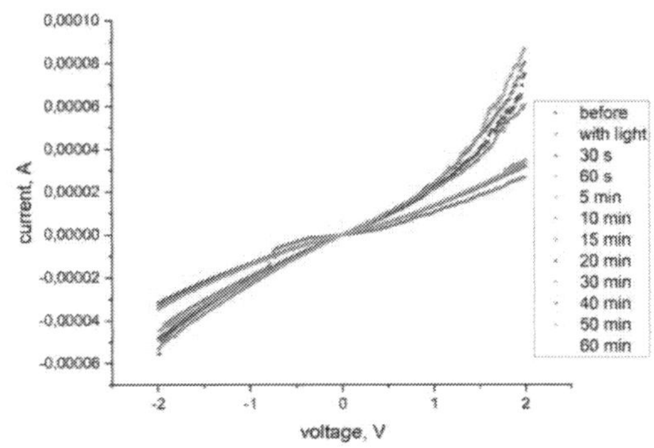

Fig. 8. The V-I family under irradiation flows of light and X-ray quanta.

An increase in current above the initial value (by about 12%) was observed, presumably due to photogeneration. It should be noted that the flux of light quanta on these samples did not lead to significant photogeneration. After exposure for 5 minutes, a sharp decrease in current (3 times) was observed, presumably due to radiation damage to the crystal structure/microstructure of the sample. During the 60 minutes during which the test was carried out, the functional heterojunction showed weakly pronounced semiconductor properties. It is concluded that the provided structures are sufficiently resistant to the effects of powerful X-ray fluxes and, probably, in the pulsed mode (micro- and millisecond pulses) they will be able to work for a long time (decades) in the form of matrix sensors of medical radiographic devices with a reduced dose load.

Since September 2021 there is an Ion Beam Analysis (IBA) active cooperation between Cuban and EG-5 researchers looking for the characterization of several functional materials developed by Cubans researchers in cooperation with colleagues of other countries, mainly from Mexico. Promoted by the EG-5 Group, in 2023 and 2024 cooperation agreements JINR-UH were signed to support this cooperation. Till December 2024 thirteen Research Problem Statements were proposed by Cubans researchers to study a variety of materials (PZT:Gd and lead free KNN ceramics, CuO/ZnO heterostructures, ZnO nanorods, Cu plasmonic nanoparticles into Cu_2O thin films, CZTS thin films growth by Isothermal Close Space Sublimation, CuO and CuO/CuO structures, Mo ground by DC and GaN, CdS, NRZnO/ZnO/ITO, CdS/ITO ground by RF Sputtering nanostructured films, TiO_2/CuO/FTO, PbS, PbS/CdS/ITO, powder from a fallen meteorite in the Viñales National Park Cuban area and zeolite material for agriculture, health and environment protection applications). Two young's Cuban researchers from the UH were trained at EG-5 in 2023. In the period 2022-2024, RBS & PIXE measurements and irradiation experiments were carried out at the ESA EG-5 to study around one hundred samples of seven Cuban groups. Since August 2024, a UH young doctorate student started to work at EG-5 for three years, and a JINR (EG-5) - UNAM (LEMA) - UH (IMRE & InSTEC) Project to *Study of zeolites for agriculture, health and environment protection applications* with several methods like IBA, AMS, NAA and D&AC, it´s carrier out. The main scientific results of this cooperation were presented in international [63-66] and national [7] conferences. Quantitative compositional characterization of $(K_{0.44}Na_{0.52}Li_{0.04})_{0.97}La_{0.01}Nb_{0.9}Ta_{0.1}O_3$ and $Pb(Zr_{0.53}Ti_{0.47})O_3$:Gd ceramics first time was reported and samples stoichiometric composition determined in two directions on each studied sample by RBS shows that powder encapsulation grown method is adequate to avoid the evaporation of K, Pb, Na elements [63-65, 67]. The structural changes and defect formation in the nanostructured CuO layer resulting from the sequential thermal treatment were studied [66].

3 ESA EG-5 Modernization

A significant part of the accelerator's capabilities cannot be realized at the moment and an urgent problem is the physical wear of the main units and installation systems and the moral aging of the scientific and experimental base. It should be noted that the problem of obsolescence and stopping of accelerators over the past 30 years in the Russian Federation was of a systemic nature. To solve this technical problem, in the framework of the JINR Seven-Year Development Plan for 2023-2030, which provides for the concentration of resources for updating the accelerator and reactor base of the Institute [68], V.N. Shvetsov, Director of the Frank Laboratory of Neutron Physics, initiated a project in 2023 to modernize the EG-5 accelerator and its experimental infrastructure.

Currently, the EG-5 accelerator is in the process of planned modernization. As part of the 03-4-1146-2-2022/2026 JINR Problem-Thematic Plan [69], the project 03-4-1146-2-2022/2026 "Modernization of the EG-5 accelerator and its experimental infrastructure" is being implemented. As a result of the implementation of the project in 2024, in cooperation with the G.I. Budker Institute of Nuclear Physics SB RAS, the parameters of the high-voltage system of the EG-5 accelerator will be significantly improved: the ion source will be replaced with a modern UHF analogue with fiber-optic control; the accelerating tube will be changed to an analogue with condenser-type ion optics; the setup for ion-beam elemental analysis of planar objects will be modernized (replacement of the high-vacuum steam-oil pump with a modern a turbomolecular analogue, installation of the setup chamber on a vibration-proof foundation, replacement of the signal processing module with a modern analogue developed at JINR, improvement of overall performance of the setup by optimizing the arrangement of the vacuum system elements).

The modernization of the high-voltage accelerator system will bring us closer to solving the main technical tasks of the project – restoring the energy range of accelerated particles: 900 keV - 4.1 MeV and increasing the ion beam current to 100-250 μA while maintaining energy stability at a level of no worse than 15 eV and its spatial stability sufficient to implement the microbeam spectrometer/nuclear microprobe option. Activities in the framework of the project in 2024 will enable us to get closer to solving the problem of creating a neutron source (energy ranges of 20 keV - 1 MeV and 3 - 5.1 MeV, fluxes of up to 10^7 particles/s cm^2), which makes it possible to study neutron-nuclear reactions (measurement of energy spectra and angular distributions of charged particles from reactions (n, α) and (n, p) / (α, n) and (p, n), integral and differential cross sections of the latter), radiation resistance of solids and conduct studies of various objects using nuclear physics methods [70, 71]. Also, in 2024, it is planned to develop complementary experimental techniques for studying the structure, microstructure, optical, electrical and electronic properties of near-surface layers of solids.

Acknowledgments. Serbia-JINR cooperation Program 2022 – 2025 (responsible from JINR A.S. Doroshkevich), Belarus-JINR cooperation Projects (responsible from JINR A.S. Doroshkevich); FLNP JINR Them 03-4-1146-1-2014/2028; T.Yu. Zelenyak, V.A. Kinev, I.O. Simonenko and A.R. Tameev acknowledge the financial support from the Russian Science Foundation (Project No. 23-19-00884, the Development of perovskite materials for radiation detection for biomedical applications); Acknowledgements, to Dr. Otilia Ana Culicov, for the support of the collaboration between Cuban researchers and the EG-5/FLNP/JINR researchers since 2021.

The authors are grateful for their help in planning the experiments and discussing the results:

Prof. R.G. Nazmitdinov, Prof. V.S. Rikhovitskiy, Prof. Yu.M. Gledenov, Prof. T. Vasilenko, Prof. A.H. Islamov, Dr. E.B. Asgerov, Dr. A.I. Madadzada, Prof. M.A. Balasoui, V.A. Alexandrov, E.A. Didenko (Joint Institute for Nuclear Research, Russia); Prof. Ya.B. Martynov (State Scientific-Production Enterprise "Istok", Russia); Prof. M. Kulik, Prof. B. Jasinska (Institute of Physics, Maria Curie-Skłodowska University, Poland); Prof. P. Badica, Prof. A. Stanculescu (National Institute of Materials Physics, Romania); Prof. M. Stef (Faculty of Physics, West University of Timisoara, Romania); Prof. L.H. Khiem, Dr. T.V. Phuc (Graduate University of Science and Technology, Vietnam; Institute of Physics, Vietnam); Prof. D. Luca ("Gheorghe Asachi" Technical University of Iasi (TUIASI), Romania); D.V Shevaldov (Dubna State University, Russia); Prof. I. Ristić (University of Novi Sad, Serbia); Dr. R. Balvanović (University of Belgrade, Serbia; National Museum Belgrade, Serbia; National Museum Požarevac, Serbia); Prof. V.F. Gremenok (State Scientific and Production Association "Scientific-Practical Center of the National Academy of Sciences of Belarus", Republic of Belarus); Prof. K.B. Bakiruly (Kazakh Research Institute of Rice Husbandry named after I.Zhakhaev, Kazakhstan).

References

1. DNP seminar, 16.10.24, Basic and applied research at the EG-5 accelerator of JINR (Dubna, Russia) and 5.5-MV Van de Graaff accelerator in Mexico in terms of development of international cooperation, (2025, January 31) from https://flnp.jinr.int/en-us/main/science/events/dnp-seminar-16-10-24.
2. Doroshkevich A., Zakharova A., Oksengendler B., Lyubchyk A., Tatarinova A., Kirillov A., Vasilenko, T.; Gorban O., Bodnarchuk V.; Nikiforova N., Zakharova E., Balasoiu M., Mardare D., Mita C., Stanculescu A., Mirzayev M., Nabiyev A., Popov E., L.H. Khiem, Donkov A., Konstantinova T. The Rectifying Contact of Hydrated Multi-Dimensional YSZ Nanoparticles for Advanced Electronics. // Nanomaterials 2022, 12, 4493. DOI: https://doi.org/10.3390/nano12244493
3. Phan Luong Tuan, Mirsolaw Kulik, Marius Stef, Tran Van Phuc, Nguyen Thi Bao My, Tatyana Yuryevna Zelenyak, Gabriel Buse, Andrei Racu, Aleksandr Doroshkevich, Le Hong Khiem, Vu Duc Cong, Andriy Igorevych Lyubchyk, Sergiy Igorevich Lyubchyk, Svitlana Borisovna Lyubchyk, Nguyen Ngoc Anh, «An examination on the porosity of ErF_3 doped CaF_2 crystal using the Rutherford back-scattering method», Nuclear Instruments and Methods in Physics Research Section B: Beam Interactions with Materials and Atoms, Volume 547, 2024, 165178, ISSN 0168-583X. DOI: https://doi.org/10.1016/j.nimb.2023.165178.
4. Aleksiayenak Yu.V., Kruglyak A.I., Bakiruly K.B., Zhalbyrov A., Baimbetova G., Tokhetova L.A., Gledenov Yu.M., Appazov N.O., Yershin Z., Doroshkevich A.S., Abstract «Utilizing ionizing radiation (gamma and neutron) to develop drought and salinity-resistant rice varieties», V International Scientific Forum «Nuclear Science and Technologies», Institute of Nuclear Physics of the Ministry of Energy of the Republic of Kazakhstan, 2024
5. T.V. Phuc, M. Kulik, D. Kołodyńska, L.H. Khiem, P.L. Tuan, J. Zuk, M. Turek,Investigations of elemental depth distribution and chemical compositions in the $TiO_2/SiO_2/Si$ structures after ion irradiation, Surface and Coatings Technology, Volume 387, 2020, 125494, ISSN 0257-8972. DOI: https://doi.org/10.1016/j.surfcoat.2020.125494.
6. Polyakov, A. Y., Vasilev, A. A., Kochkova, A. I., Shchemerov, I. V., Yakimov, E. B., Miakonkikh, A. V., Chernykh, A. V., Lagov, P. B., Pavlov, Y. S., Doroshkevich, A. S., Isaev, R. S., Romanov, A. A., Alexanyan, L. A., Matros, N., Azarov, A., Kuznetsov, A., &Pearton, S. (2024). Proton damage effects in double polymorph γ/β-Ga_2O_3 diodes. Journal of Materials Chemistry C, 12(3), 1020–1029. DOI: https://doi.org/10.1039/D3TC04171A
7. Tuan P.L., Kulik M., Turek M., Duc P.M., Phuc T.V., Anh N.N., Cong V.D., Bao My N.T., Doroshkevich A.S. Examination of the Native Oxide Layer Using the RBS/NR Method on the Surface of GaAs Samples Before and After Hot-Implanted Al Ions, ISINN-30 proceedings.
8. P.L. Tuan, M. Kulik, T.V. Phuc, A.I. Madadzada, T.Yu. Zelenyak, M. Turek, J. Zuk, C. Mita, A. Stanculescu, A.S. Doroshkevich, B. Jasinska, L.H. Khiem, N.N. Anh, N.T. Bao My, Pseudo-dielectric function spectra of the near surface layer of GaAs implanted with various fluence of Xe^+ ions, Thin Solid Films, Volume 756, 31 (2022), 139376. DOI: https://doi.org/10.1016/j.tsf.2022.139376.
9. Horodek P., Kulik M., Application of positron beam for the long range effect studies in proton implanted iron, Nuclear Inst. and Methods in Physics Research B.443 (2019) 84–89. DOI: https://doi.org/10.1016/j.nimb.2019.02.003
10. Yu. M. Gledenov, M. V. Sedysheva, G. Khuukhenkhuu and el. Measurement of the cross sections of the $^{25}Mg(n,\alpha)^{22}Ne$ reaction in the 4–6 MeV region, Phys. Rev. C, 98, 034605 (2018). DOI: https://doi.org/10.1103/PhysRevC.98.034605
11. Jie Liu, Zengqi Cui, Yiwei Hu, Haofan Bai, Yi Yang, Xichao Ruan, Cong Xia, Jinxiang Chen, Guohui Zhang, Yu M Gledenov, E Sansarbayar, G Khuukhenkhuu, L Krupa, I Chuprakov, Hanxiong Huang, Jie Ren, Qiwen Fan, Yuan Gao and Xiangjun Yang, $^{63}Cu(n,\alpha)^{60}Co$ cross sections in the MeV region, J. Phys. G: Nucl. Part. Phys., Vol. 50, (2023). DOI: https://doi.org/10.1088/1361-6471/acb960
12. E. Sansarbayar, Yu. M. Gledenov, I. Chuprakov, G. Khuukhenkhuu, G. S. Ahmadov, L. Krupa, Guohui Zhang, Haoyu Jiang, Zengqi Cui, Yiwei Hu, Jie Liu, N. Battsooj, I. Wilhelm, M. Solar, R. Sykora, and Z. Kohout, Cross sections for the $^{35}Cl(n,\alpha)^{32}P$ reaction in the 3.3–5.3 MeV neutron energy region, Phys. Rev. C 105, 049902 (2022). DOI: https://doi.org/10.1103/PhysRevC.104.044620
13. Uglov V.V. Methods of analysis of elemental composition of surface layers, BSU, 2007.
14. Oura K., Lifshits V. G., Saranin A. A., et al. Introduction to Surface Physics, Nauka, 2006.
15. Komarov F.F., Kumakho M.A., Tashlykov I.S. Non-destructive analysis of solid surfaces by ion beams, Universitetskoe, 1986.
16. Tashlykova-Bushkevich I.I. The Rutherford backscattering method in the analysis of solids. BSUIR, 2003.
17. Zelenyak T.Yu. et al. Effect of Cs ion implantation on the atomic composition and structure of functional architectures: glass/ITO/$TiO_2/CH_3NH_3PbI_3$. IV Moscow Autumn International Conference on Perovskite Photovoltaics (MAPPIC-2022), September 19-21, 2022, Moscow, Russia.
18. Kruglyak A.I., Soloviev A.A., Alekseenok Yu.V., Doroshkevich A.S. Induction of variability in triticale (× triticosecale) using fast neutrons // Abstracts of reports of the V International Scientific

Forum "Nuclear Science and Technologies" (15th International Conference "Nuclear and Radiation Physics", 4d International Conference "Nuclear and Radiation Technologies in Medicine, Industry and Agriculture"), Republic of Kazakhstan, 2024.

19. Siemek, K., Kulik, M., Eseev, M. et al. Surface and Subsurface Defects Studies of Dental Alloys Exposed to Sandblasting. Acta Metall. Sin. (Engl. Lett.) 32, 1181–1194 (2019). DOI: https://doi.org/10.1007/s40195-019-00884-5
20. A. Olejniczak, N.A. Nebogatikova, A.V. Frolov, M. Kulik, I.V. Antonova,V.A. Skuratov, Swift heavy-ion irradiation of graphene oxide: Localized reduction and formation of sp-hybridized carbon chains, Carbon, Volume 141, 2019,Pages 390-399,ISSN 0008-6223, (2018), DOI: https://doi.org/10.1016/j.carbon.2018.09.042
21. Phuc T.V., Kulik M., Kobzev A.P., Khiem L.H., Study of Elemental Depth Distribution in The Material $TiO_2/SiO_2/Si$ by Rutherford Backscattering Spectrometry (RBS), Communications in Physics, 29, 3SI (2019), pp. 393-400. DOI: https://doi.org/10.15625/0868-3166/29/3SI/14328.
22. Tran Van Phuc, Miroslaw Kulik, Le Hong Khiem, Afag Madadzada, Marcin Turek, Dorota Kołodyńska, Phan Luong Tuan, Nguyen Ngoc Anh, Mai Quynh Anh, Nguyen Van Tiep, Krzysztof Siemek, Variation of TiO_2/SiO_2 mixed layers induced by Xe^+ ion irradiation with energies from 100 to 250 keV, Materials Science and Engineering: B,Volume 277, 2022, 115566, ISSN 0921-5107. DOI: https://doi.org/10.1016/j.mseb.2021.115566.
23. A.Y. Polyakov, V.I. Nikolaev, A.I. Pechnikov, P.B. Lagov, I.V. Shchemerov, A.A. Vasilev, A.V. Chernykh, A.I. Kochkova, L. Guzilova, Yu.S. Pavlov, T.V. Kulevoy, A.S. Doroshkevich, R.Sh. Isaev, A.V. Panichkin and S.J. Pearton, Carrier removal rates in 1.1 MeV proton irradiated α-Ga_2O_3 (Sn), 2023 J. Phys. D: Appl. Phys. 56 305103 DOI: https://doi.org/10.1088/1361-6463/acd06b
24. Ivan Schemerov, A.Ya. Polyakov, P. B. Lagov, V. D. Kirilov, Svetlana Kobeleva, A. I. Kochkova, Yu. O. Kulanchikov, O. S. Doroshkevich. The effect of trapping sites introduced by 1 MeV proton irradiation on the reverse current recovery time in Ga_2O_3-based Schottky diodes. Industrial laboratory Diagnostics of materials, 89, 7, 2023. DOI: https://doi.org/10.26896/1028-6861-2023-89-7-25-33
25. B.R. Kutlimurotov, A.S. Doroshkevich, A.I. Lyubchyk, B.L. Oksengendler, N.N. Nikiforova, M. Adilov, R.Kh. Ashurov, S.X. Suleymanov, A.S. Zakharova, E.A. Gridina, C. Mita, The variance of the electronic structure in the near-surface regions of chemically homogeneous nanoparticles of oxide materials and its role in the conversion of chemisorption energy of water on the powder of zirconium dioxide, Uzbek Journal of Physics, Vol. 24, No. 4, pp. 254-262, 2022. DOI: https://doi.org/10.52304/.v24i4.378
26. K.L. Keldysh, Soviet Physics JETF, 1964.
27. N. N. Nikiforova, B.L. Oksengendler et al., Fractal Nanocatalysis and Related Phenomena, Journal of Surface Investigation: X-ray, Synchrotron and Neutron Techniques, 2018, № 2, c. 68–73. DOI: https://doi.org/10.7868/S0207352818020117
28. Patent for invention "Solid-state capacitor ionistor with a dielectric layer which is made from a dielectric nanopowder" Rospatent no. 2729880 Authors Doroshkevich A. S., Shilo A.V., Zelenyak T. Yu., Konstantinova T. E., Lyubchik A.V., Tatarinovaa. A., Gridina E. A., Doroshkevich N. V. Patent Owner: JINR. Application no. 2019135580; priority of the invention 5.11.2019; date of state registration in the state register of inventions of the Russian Federation 13.09.2020, term of the exclusive right-2039.
29. Doroshkevich Alexander S., Lyubchyk Andriy I., Oksengendler Boris L., Zelenyak Tatyana Yu., Appazov Nurbol O., Kirillov Andriy K., Vasilenko Tatyana A., Tatarinova Alisa A., Gorban Oksana O., Bodnarchuk Viktor I., Nikiforova Nadejda N., Balasoiu Maria, Mardare Diana M., Mita Carmen, Dorin Luca, Mirzayev Matlab N., Nabiyev Asif A., Popov Evgeni P., Stanculescu Anca I., Konstantinova Tatyana E., Aleksiayenak Yulia V. The effect of electric energy accumulation by hydrated ZrO_2 – nanoparticles // Nanomaterials 2022, 12, 1783. DOI: https://doi.org/10.3390/nano12111783
30. Artem Shylo, Igor Danilenko, Oksana Gorban, Oleksandr Doroshkevich, Igor Nosolev, Tetyana Konstantinova, Andriy Lyubchyk Hydrated zirconia nanoparticles as media for electrical charge accumulation // Journal of Nanoparticle Research J Nanopart Res (2022) 24:18. DOI: https://doi.org/10.1007/s11051-022-05407-5.
31. Nematov D.D., Burkhonzoda A.S., Khusenov M.A., Kholmurodov K.T., Doroshkevich A.S., Doroshkevich N.V., Zelenyak T.Yu., Majumder S. Molecular Dynamics of DNA Damage and Conformational Behavior on a Zirconium-Dioxide Surface. Journal of Surface Investigation: X-ray, Synchrotron and Neutron Techniques. 2019;13(6):1165-1184. DOI: https://doi.org/10.1134/S1027451019060430
32. Khusenov Mirzoaziz, Nematov Dilshod, Burhonzoda Amondullo, Kholmurodov Kholmirzo, Doroshkevych Aleksandr, Doroshkevych Nelly, Zelenyak Tatyana and Majumder Subrata. Molecular Dynamic Studies of the DNA Radiation and Conformation Processes on a Zirconium Dioxide Surface, J. Rad. Nucl. Appl. 4, No. 3, 157-164 (2019). DOI: https://doi.org/10.18576/jrna/040304
33. Doroshkevich A.S., Shylo A.V., Glazunova V.A., Volkova G.K., Kirillov A.K., Zelenyak T. Yu., Burkhovetskiy V.V., Turchenko V.A., Doroshkevich V.S., Nabiyev A.A., Vasylenko T.A.,

Islamov A. Kh., Craus M.L. Self-organization processes in nanopowder dispersed system based on zirconia under pressure action, Results in Physics. 16 (2020) 102809. DOI: https://doi.org/10.1016/j.rinp.2019.102809

34. Lakusta Marharyta, Danilenko Igor, Volkova Galina, Loladze Larisa, Burkhovetskiy Valeriy, Doroshkevich Oleksandr, Brykhanova Irina, Popova Inna, Konstantinova Tetyana. Sintering kinetics of ZrO_2 nanopowders modified by group IV elements. 2019, International Journal of Applied Ceramic Technology. DOI: https://doi.org/10.1111/ijac.13215
35. Bacherikov Yuriy Yu., Lytvyn Peter M., Okhrimenko Olga B., Zhuk Anton G., Kurichka Roman V., Doroshkevich Aleksandr S. Surface potential of meso-dimensional ZnS:Mn particles, obtained using SHS method. JNanopartRes. (2018) 20, 316. DOI: https://doi.org/10.1007/s11051-018-4413-1
36. Doroshkevich A.S., Nabiev A.A., Shylo A.V., Pawlukojć A., Doroshkevich V.S., Glazunova V.A., Zelenyak T.Yu., Doroshkevich N.V., Rahmonov Kh.R., Khamzin E.Kh., Nematov D.D., Burhonzoda A.S., Khusenov M.A., Kholmurodov Kh.T., Majumder S., Balasoiu M., Madadzada A., Bodnarchuk V.I., Frequency modulation of the Raman spectrum at the interface DNA - ZrO_2 Nanoparticles, Egyptian Journal of Chemistry, 2019. DOI: https://doi.org/10.21608/ejchem.2019.12898.1806
37. Gridina Elizaveta, Doroshkevich Aleksandr, Lyubchyk Andriy, Shylo Artem, Asgerov Elmar, Madadzada Afag, Zelenyak Tatyana, Balasoiu Maria, Lazar Diana, Orelovich Oleg. The effect of percolation electrical properties in hydrated nanocomposite systems based on polymer sodium alginate with a filler in the form nanoparticles ZrO_2-3mol% Y_2O_3, Advanced Physical Research, 1(2), 70-80.
38. Gladyshev P.P., Banavoth M., Swetha T., Bingwa N., Martynov Ya. B., Zelenyak T. Yu., Kinev V.A. and Nazmitdinov R.G. Hybrid Organo-Inorganic Perovskite Solar Cells: Architecture Evolution, Materials of Functional Layers, Photoelectric Characteristics, Properties, and Effciency. Book Perovskite solar cells. 2019. P. 1-76.
39. Magon N.S., Zelenyak T.Yu., Korolik O.V., Gladyshev P.P., Mazanik A.V. Effect of pyridine treatment on the optical properties of organo-inorganic perovskite films. Journal of the Belarusian State University. Physics, 2019, 66–72.
40. Zelenyak T.Yu., Kinev V.A., Fokina S.I., Gladyshev P.P., Ibrahim M.A. Optical Properties of Thin-film Hybrid Organic-inorganic Perovskites under The Influence of Pyridine Vapor. Egypt. J. Chem., 2019, 62, 69-75.
41. Zelenyak T., Kinev V., Rezepov P., Korolik O., Mazanik A., Tivanov M., Gladyshev P. Formation and optical properties of hybrid organic-inorganic $MAPbI_3$ perovskite films. IOP Conference Series: Materials Science and Engineering. 2019. V. 498 P. 012012. DOI: https://doi.org/10.1088/1757-899x/498/1/012012
42. Zelenyak T.Yu., Rakhmonov Kh.R., Nematov D.D., Burkhonzoda A.S., Khusenov M.A., Kholmurodov Kh.T.. Molecular-dynamics modeling of structural stability of perovskite materials as elements of solar cells. The scientific-practical international conference of students, magisters, Phd students and young scientists «MUHANDIS-2019» TTU, Dushanbe, March 14 – 16, 2019. P. 8-21.
43. A.I. Kruglyak, A.S. Doroshkevich, Yu. Aleksiayenak, N.O. Appazov, K.B. Bakiruly, M. Balasoui, M.N. Mirzayev, A.A. Nabiyev, E. Popov, Application of the accelerator mass spectrometry method to study the mechanisms of radiation mutagenesis of rice crops: the current state of the issue, Proceedings of the International Conference: Condensed Matter Research at the IBR-2 (Dubna, JINR Apr. 25–29, 2022) 65-66.
44. D.A. Komarova, A.S. Sokhatsky, A.I. Krylov, N.V. Kurylev, S.V. Mitrofanov. The principle of homogeneous ion doping of material samples for structural studies. Interaction of radiation with a solid: Proc. of the 14th Int. Conf. dedicated to the 100th anniversary of Belorussian State University, Minsk, Belarus, September 21–24, 2021.
45. Polyakov A. Y. et al. Carrier removal rates in 1.1 MeV proton irradiated α-Ga_2O_3 (Sn), Journal of Physics D: Applied Physics. 2023, 56, 30. 305103. DOI: https://doi.org/10.1088/1361-6463/acd06b
46. Utamuradova Sh.B., Daliev Sh.Kh., Rakhmanov D.A., Doroshkevich A.S., Kinev V.A., Ponomareva O.Yu., Mirzayev M.N., Isayev R.Sh. (2023). IR – spectroscopy of n-Si<Pt> irradiated with protons. Advanced Physical Research, 5(2), 73-80.
47. Isaev R.Sh., Safonov D.A., Dzhumaev P.S., Korenevsky E.L. Features of high-temperature oxidation of zirconium alloys with a chromium coating obtained by magnetron sputtering, Non-ferrous metals, (2022), 10. DOI: https://doi.org/10.17580/tsm.2022.10.04
48. A. Vladescu (Dragomir) et al. Effect of Si and Nb additions on carbonitride coatings under proton irradiation: A comprehensive analysis of structural, mechanical, corrosion, and neutron activation properties, Nuclear Materials and Energy. 2023. P. 101457. DOI: https://doi.org/10.1016/j.nme.2023.101457
49. D. Neov, A.I. Beskrovnyi, A.S. Abiyev, D.M. Mirzayeva, E. Demir, A.H. Valizade, T.T. Hlatshwayo, T.T. Thabethe, O.A. Samedov, R.N. Mehdiyeva, K. Hasanov, Z.A. Sharipov, A.S. Doroskevich, F. Mamedov. Effects of helium ion irradiation on tungsten-based composites: Neutron diffraction; as a realtime diffractometer. Advanced Physical Research, 2023, 5(2), 95-102.
50. Isayev R. et al. Corrosion of EP823 Steel Cladding Under Heavy Liquid-Metal-Coolant Reactor Conditions: A Review, Nuclear Science and

Engineering, 2024, 1-19. DOI: https://doi.org/10.1080/00295639.2024.2383110

51. R.Sh. Isayev, P.S. Dzhumaev, I.A. Naumenko, M. V. Leontieva-Smirnova, Corrosion resistance of chromium coating on the inner surface of EP823-Sh steel cladding, Nuclear Energy and Technology, 2024, 10, №. 2. C. 81-88. DOI: https://doi.org/10.3897/nucet.10.119642/
52. R.Sh. Isaev, P.S. Dzhumaev "Interaction of chromium coating with fuel element cladding made of EP823-Sh steel in the temperature range of 420-650 °C, Bulletin of NRNU MEPhI, 2024,13, 4, 273-281.
53. D.R. Belichko, G. K. Volkova, A. V. Maletsky, R. Sh. Isaev, Effect of proton irradiation on the structure and properties of composite ceramics of the composition YSZ–SiO_2–Al_2O_3, Materials Science Issues, 2024, No. 3 (119).
54. A.V. Maletskii, G.K. Volkova, D.R. Belichko, V.A. Glazunova, A.S. Doroshkevich, A.A. Tatarinova, S.I. Lyubchyk, S.B. Lyubchyk, Influence of stabilized zirconium dioxide and high hydrostatic pressure on the kinetics of sintering nanopowders of metastable aluminum oxide, Ceramics International, 2024, ISSN 0272-8842, DOI: https://doi.org/10.1016/j.ceramint.2024.09.002
55. A.V. Maletskyi, T.E. Konstantinova, G.K. Volkova, D.R. Belichko, A.S. Doroshkevich, E. Popov, N. Cornei, B. Jasinska, Zh.V. Mezentseva, A.A. Tatarinova, M.N. Mirzayev, L.H. Khiem, I. Ristić, V. Teofilović, R. Balvanović, High hydrostatic pressure influence on the properties and tendency to agglomeration of ZrO_2 grains of the Al_2O_3 – YSZ composite ceramics system, Ceramics International, 2023, https://doi.org/10.1016/j.ceramint.2023.01.202
56. D. Xu, Z. Yao, J. Pan, H. Feng et al., Study on the multiple characteristics of M3 generation of pea mutants obtained by neutron irradiation. Nucl. Sci. Tech. 31, 67 (2020). https://doi.org/10.1007/s41365-020-00777-8
57. W. Zhang, L. Jiao, M. Hoshi, Study of micronucleus induction in the root-tip cells of onion seedlings after irradiation as dry dormant seeds by 252Cf fission neutrons. J. Radiat. Res. Radiat. Process. 23(1), 15–18 (2005).

    https://doi.org/10.3969/j.issn.1000-3436.2005.01.004
58. Mei, L. (2010). Effect of fast neutron irradiation on amylose content and amylopectin structure of Thailand rice cultivar Jao Hom Nin. Agric. Sci. Technol. 11, 19–21.
59. Ruengphayak, S., Ruanjaichon, V., Saensuk, C., Phromphan, S., Tragoonrung, S., Kongkachuichai, R., et al. (2015). Forward screening for seedling tolerance to Fe toxicity reveals a polymorphic mutation in ferric chelate reductase in rice. Rice 8, 3. https://doi.org/10.1186/s12284-014-0036-z
60. Bakiruly, K.; Zhalbyrov, A.; Kruglyak, A.; Aleksiayenak, Y.; Baimbetova, G.; Yershin, Z.; Gledenov, Y.; Doroshkevich, A.; Appazov, N. Creation of Salinity and Drought Resistant Mutant Rice Forms by Ionizing Radiation (Gamma and Neutron Radiation). *Preprints* 2023, 2023040590. DOI: https://doi.org/10.20944/preprints202304.0590.v1
61. K. Bakiruly; A.I. Kruglyak; Z.R. Yershin; A. Zhalbyrov; G. Baimbetova; N.O. Appazov Use of ionizing radiation for obtaining mutant forms resistant to salinity and drough factors for their application in synthetic rice breeding (in Russian) Bulletin of the Korkyt Ata Kyzylorda University, №3-1 (66), pp. 55-65, 2023 DOI: https://doi.org/10.52081/bkaku.2023.v66.i3.067
62. Kruglyak A.I., Soloviev A.A., Alekseenok Yu.V., Doroshkevich A.A. Induction of variability in triticale (× triticosecale) using fast neutrons // Abstracts of reports of the V International Scientific Forum "Nuclear Science and Technologies" (15th International Conference "Nuclear and Radiation Physics", 4d International Conference "Nuclear and Radiation Technologies in Medicine, Industry and Agriculture"), Republic of Kazakhstan, 2024. – p. 297.
63. J. Pérez Pérez, P.L. Tuan, T.Yu. Zelenyak, A.I. Kruglyak, A.S. Doroskevich, M.S. Pomares Alfonso, O. Díaz Rizo, G. Pérez Zayas, J. Portelles Rodríguez, J. Fuentes Betancourt, J.M. Siqueiros, R. López, Y. de Armas, J. Rebellón-Watson, A. Leyva Fabelo, E. Pedrero González. Qualitative analysis of KNN and PZT ceramics by Particle Induced X-Ray Emission and Rutherford Backscattering Spectrometry. XVIII Workshop on Nuclear Physics (WONP-2022), October 17 to 21, 2022; Institute of Technologies and Applied Sciences (InSTEC) of the University of Havana, Havana, Cuba.
64. R. Méndez, P.L. Tuan, A. Kruglyak, J. Pérez Pérez, J. Portelles, T.Ye. Konstantinova, O. Díaz, G. Perez, M. Pomares, E. Pedrero, A.S. Doroshkevich. Results of the characterizations of samples of Cuban materials using nuclear and related techniques available at the Frank Laboratory of Neutron Physics, JINR. AYSS-2023, October 29 to November 3, 2023. Dubna, Russia.
65. S. Fortuné Fábregas, T.L. Phan, A.S. Doroshkevich, E. Pedrero González, A. Iribarren Alfonso, L. Vaillant Roca. Study of CuO nanostructures modified with a sequential thermal treatment for photovoltaic applications. AYSS-2024, 28 October to 1 November, 2024. Dubna, Russia.
66. A.S. Doroshkevich, P.L. Tuan, T.Yu. Zelenyak, A.I. Kruglyak, T.Ye. Konstantinova, J. Portelles Rodriguez, J.E. Fuentes Betancourt, A. Leyva Fabelo, J. Pérez-Pérez, S. Fortuné-Fábregas, R. Méndez-Hernández, J. Alba Cabañas, L. Vaillant-Roca, S. Larramendi Valdés, A. Fundora Cruz, A. Iribarren Alfonso, F. Forcade Zamora, G. Rodríguez-Fuentes, J.C. Molina Amat, G. Pérez-Zayas, M.S. Pomares-Alfonso, O. Díaz Rizo, M. Díaz Castro, E. Pedrero-González. Progress of the

collaboration between the University of Havana and the EG-5/FLNP/JINR Group in materials characterization. XIX Workshop on Nuclear Physics (WONP-2024); October 7 to 11, 2024; Institute of Technologies and Applied Sciences (InSTEC), University of Havana, Havana, Cuba.

67. R. Méndez-Hernández, J. Pérez Pérez, P.L. Tuan, T.Yu. Zelenyak, A.I. Kruglyak, J. Portelles Rodríguez, J. Fuentes Betancourt, J.M. Siqueiros, R. López, Y. de Armas, J. Rebellón-Watson, A. Leyva Fabelo, T.Ye. Konstantinova, A.S. Doroskevich, E. Pedrero González. Determination by Rutherford Backscattering Spectrometry of the stoichiometric distribution of $KNNLa_{0.01}$ and PZT:Gd ceramics grown by powder encapsulation method. XVI Simposio de la Sociedad Cubana de Física, 8 - 11 abril 2024, Universidad de La Habana, Cuba.
68. JINR Basic facilities, (2025, January 31) from https://www.jinr.ru/jinr_facilities-en/
69. Topical Plan for JINR Research and International Cooperation 2025. URL: https://www.jinr.ru/wp-content/uploads/JINR_Docs/JINR_Topical_Plan_2025%20(eng).pdf
70. Komar E. G., Fundamentals of Acceleration Technology, Moscow, 1975; Accelerators. Collection, edited by B. N. Yablokov, Moscow, 1962.
71. Electrostatic accelerators of charged particles. Collection, edited by A.K. Walter, Moscow, 1963.

Evaluating the possible degradation of extremely thin silicon detectors within a gas jet target environment

Fabiola Silva[1,*], *David* Godos[1,2], *Libertad* Barrón[1], *Erika* Ruíz[1], *José* Gómez[1], *Óscar* López[1], *Frida* Quintero[1], *Guadalupe* Reza[1], *Roberto* Gleason[1], *Carlos* Valencia[1], and *Luis* Acosta[3,1,**]

[1]Instituto de Física, Universidad Nacional Autónoma de México, 04510, Mexico City, Mexico
[2]Facultad de Ciencias Experimentales, Universidad de Huelva, 21071, Huelva, Spain
[3]Instituto de Estructura de la Materia, Consejo Superior de Investigaciones Científicas, 28006 Madrid, Spain.

Abstract. The SUGAR system (SUpersoni GAs jet-taRget) is a device which may reach an ultrathin and pure gas target. The main idea of SUGAR is to use a particle detection array inside its scattering chamber in order to study interesting reactions at low energy. In this direction, it is essential to evaluate the behaviour of thin silicon detectors surrounding the jet while it circulates, considering the turbulence expected due to its supersonic nature. One of the variables to control the possible damage of the silicon surface is the resolution, this should get worse if the turbulent gas is causing some effect on it.
With this in mind, in this work we present the results obtained from the analysis of three different types of silicon detectors: a thick Passivated Implanted Planar silicon detector, a extremely thin surface barrier detector and a very thin Double-Sided Silicon Strip Detector. Such devices were tested for a number of hours inside the jet environment, controlling every time the detector behaviour, by using radioactive sources and their spectroscopic data. The results obtained are a first glimpse about the very acceptable possibilities to use thin detectors, thin strip detectors and telescopes inside the SUGAR scattering chamber.

1 Introduction

The system, called SUGAR (SUpersonic GAs jet-taRget), is a complex mechanical system that allows maintaining a differential vacuum contact between a beam line and a chamber (jet chamber). Inside this chamber, a supersonic gas jet circulates, producing a “pure jet target” that resembles a curtain of particles. The gas continuously enters and leaves the jet chamber thanks to a robust pumping system. SUGAR also has a differential vacuum that provides the necessary conditions to maintain the jet chamber, also used as scattering chamber, in a windowless mode. Being gaseous and maintained in continuous circulation, the supersonic gas jet target, is ultrathin (around of 10^{18} particles/cm^2 thickness), so that it allows measuring particle scattering towards forward angles, even at low energies, such as those used by all research accelerators existing in Mexico. A detailed description of the SUGAR system can

*e-mail: fabiolasilva@estudiantes.fisica.unam.mx
**e-mail: luis.acosta@csic.es

be consulted on Ref. [1]. A layout of SUGAR, where its elements are indicated, is shown in Fig. 1a, basically, the three main components: the jet chamber, which also works as scattering chamber; the differential pumping system, which allows the windowless connection with the beam line; and the jet pumping system, which generates the jet by using a group of 4 specific pumps. The Fig. 1b shows the interior of the scattering chamber, where the upper part is a pipe which works as "the nozzle" to inject gas inside the chamber; a central pipe (ISO-K-100, "the catcher") is directly connected to a high speed mechanical pump which exhaust most of the entering gas; the gas spilling out the catcher is pumped out from the bottom by a large high speed roots pumps system, which is connected to the larger chamber output (ISO-K-250). The system is completed with a metallic disk screwed to the catcher pipe, where a silicon detection array can be placed [1].

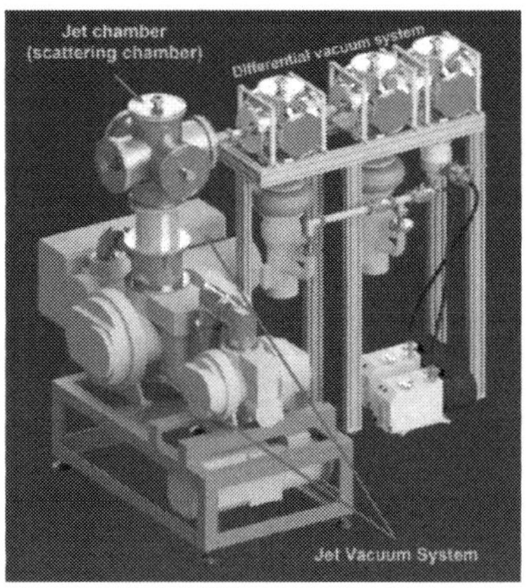

(a)

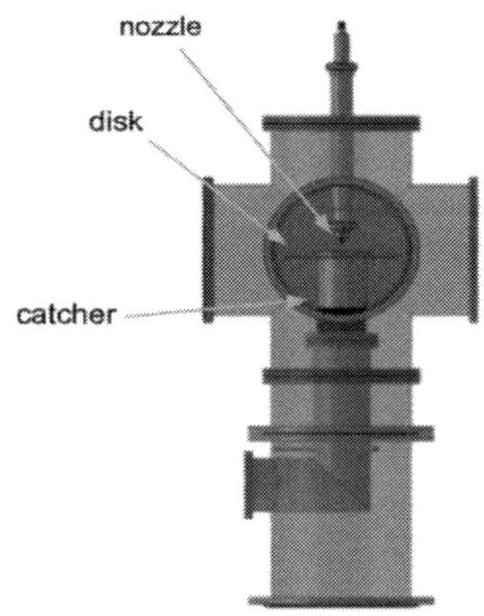

(b)

Figure 1: a) Main components of the SUGAR system. b) Components inside the jet/scattering chamber (figure modified from [1]).

A supersonic gas is the best approximation to an "ideal" target and it is similar to a solid target. Being supersonic, the particles composing the gas will travel at speeds greater than the speed of sound, thus creating a thin curtain at the atomic level of the gas used (see the simulation made with the COMSOL software [2] of the Fig. 2). To produce the supersonic gas jet target, it is necessary to inject high-pressure gas (between 1-5 atm) into the scattering chamber, where a low vacuum of 10^{-1} torr is permanently reached while the gas is being injected. The very thing aperture in the nozzle forces the production of a supersonic flow. The simulation of the Fig. 2 shows a range of up to around Mach 4 in the central zone (red), dropping to Mach 1.5 at 7.5 mm from the gas outlet (where the green color begins). The most intense zone, which is centered in the beam, is 5 mm x 3 mm in area, according to the simulation.
It is important to mention that turbulences generated by the jet inside the scattering chamber of SUGAR could affect the behaviour of thin silicon detectors as well as damage their structure. For example, turbulences can generate vibrations, which in turn increases electronic noise. This noise can make it difficult to distinguish between two individual events generated by the incident radiation, thus reducing the energy resolution of the detector.

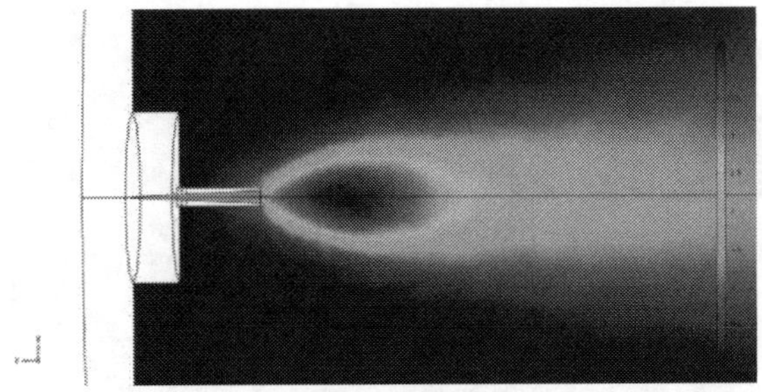

Figure 2: Simulation of the jet shape, using the standard parameters of SUGAR system. Simulation was produced with the COMSOL software. The color scale indicates 'Mach' units, such that 1 = 343.2 m/s.

2 Experimental Setup

The measurements were carried out in the 5.5 MeV Van der Graaff accelerator hall of the Carlos Graef Fernández Laboratory of the Instituto de Física, UNAM (IFUNAM), were SUGAR is presently mounted. In order to evaluate the possible damage of silicon detectors, particularly very thin ones (those that are useful in low energy reaction to built telescopes), we tested different types of silicon detectors inside of the SUGAR scattering chamber: A thick Passivated Implanted Planar Silicon detector (PIPS) of 500 μm, a thinner one, a Surface Silicon Barrier Detector (SSBD) of 12 μm, to form a ΔE-E telescope with them. Both detectors have an active area of around of 1 cm^2. As second test we used a thin 16 $\times$ 16 Double-Sided Silicon Strip Detector (DSSSD) of 60 μm thick, in a similar condition as the telescope. With the DSSSD it is possible to make a hitmap combining the signal of the back and forward strips forming pixels. The bidimensional ΔE-E$_{tot}$ as well as the matrix formed with the hitmaps could give us quantitative information regarding the possible loose of the detection resolution, which would mean a deterioration of the thin detectors by the action of the jet turbulences [3–6].
In order to simulate a particle beam, we used a monochromatic alpha source of ^{241}Am. The source (taken from the source storage of the IFUNAM) was sealed several years ago, so the peak is slightly spread, resulting worse resolution. Then, the resolution was better controlled with a standard triple alpha source, composed by ^{239}Pu, ^{241}Am and ^{244}Cm. A 3 cm long 3D printed plastic cannon was fixed in front of the radioactive source to better reproducing a particle beam. For the second part of the experiment, we used the DSSSD in front of the monochromatic alpha source. The jet target is always in the middle of the alpha source and the respective detection system, in order to control every time the alphas crossing the jet and arriving to the detectors. The two experimental setups used for this study are shown in the Figs. 3a and 3b.

The readout of detectors was made by using Mesytec MPR modules [7] (MPR-1 and MPR 32 for telescope and DSSSD, respectively); signals were digitized with FEBEX-3 data adquisition system [8] in a self-trigger mode. The digitizers work normally with the preamplifier signal, which could be noisy in a self-trigger mode. For this reason we tested the resolution of single detectors with an analogical CAMAC ADC [9] and filtering with an ORTEC 512 amplifier [10] and a GATE generated with it. The maximum real resolution of the E detector can be seen in the spectrum of Fig. 4a with a Full Width at Half Maximum (FWHM) of 23 keV. Nevertheless, we decided to use for the whole measurement the digitizer system, considering that FEBEX-3 cards are very flexible to manage the signals for strip detectors

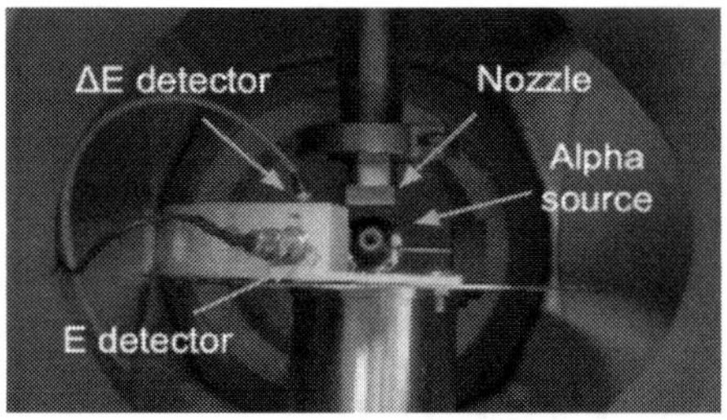

(a)

(b)

Figure 3: Images corresponding to the setups from the two parts of the experiment a) for the telescope ΔE-E case and b) DSSSD detector case. See the text for details.

(they have 16 channels single end inputs each). For the final measurement we also decided to remove the alpha source cannon, in order to increase the rate of alphas in all detectors. With these decisions, the FWHM decreased to 93 KeV for the E detector, and resulting in 52 keV (front side) and 85 keV (back side) for the DSSSD. The change in the E detector resolution can be well observed in the Fig. 4b, comparing with Fig. 4a case; one of the triple alpha spectra obtained for calibration in case of one front strip of the DSSSD can be seen in Fig. 4c. A summary of the different resolutions obtained is presented in Table 1.

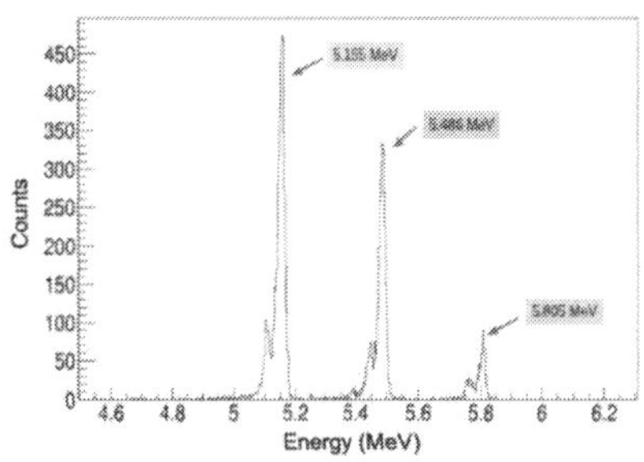

(a)

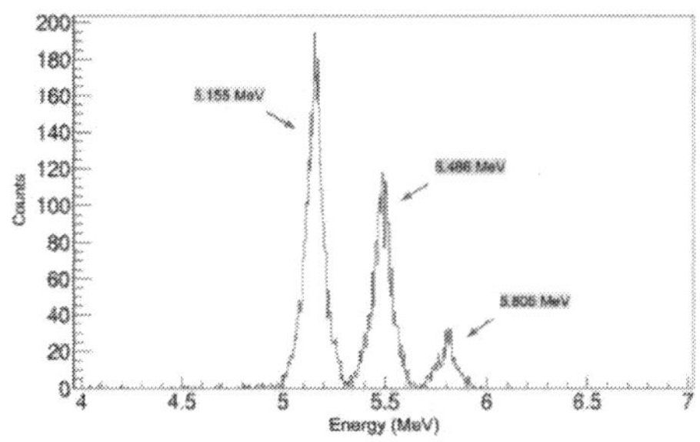

(b)

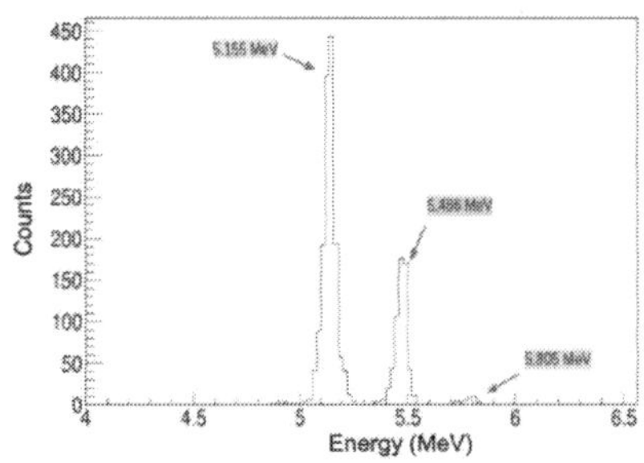

(c)

Figure 4: Calibrated spectra for E (PIPS detector) and DSSSD detectors with different data adquisition systems.

Even with this loss of resolution, the signal in detectors was well controlled, considering the important goal was to know the initial resolution in order to be sure whether it shows

https://doi.org/10.1051/epjconf/202533303003

Table 1: Results of the resolution measurements for the E detector and 2 strips of DSSSD (front and back), by using a triple alpha source: FWHM and Resolution are shown for every detector; Ho, is the center of the alpha peak; μ is the mean of the data collection, σ is the standard deviation, and δR is the resolution error.

Detector	FWHM (MeV)	Ho (MeV)	μ (MeV)	σ (MeV)	Resolution (%)	δ R (%)
E detector (CAMAC)	0.023	5.155	5.155	0.009	0.44	±0.01
E detector (FEBEX)	0.093	5.155	5.155	0.039	1.80	±0.02
Strip 0 Front DSSSD	0.052	5.141	5.141	0.022	1.01	±0.01
Strip 2 Back DSSSD	0.085	5.132	5.132	0.036	1.65	±0.01

changes with respect to the time of measurement. Regarding the possible scattering of alpha particles and jet: the low rate of particles from the radioactive source and the thin jet target, make practically no possible any interaction between them (this situation was proved with a simulation in SRIM [11] including the alpha source rate and the composition and target thickness). This condition was not an issue considering the study was centered in the interaction between jet and detector, therefore we could use the counting of alphas as parameter to monitor the detector resolution. Measurements with jet switched ON and OFF were made, using periods of 8 hours of data taking. The obtained results are described in the following section.

3 Results and Analysis

In order to evaluate the behaviour of the three different types of silicon detectors when exposed to the action of the jet, data collection was taken during 8 hours shifts without jet and with jet for the case of ΔE-E telescope and with jet only for the case of DSSSD. With this, the energy spectra and FWHM values were obtained. For the first case, when the telescope was tested, we generated spectra for every two hours of data taken. Fig. 5a shows the superposed spectra when the jet was not working; on the other hand, the Fig. 5b shows the superposed spectra obtained when the jet was working. Comparing both groups of spectra, there is not any observable change in resolution at first glance. In order to prove quantitatively this fact we generated integration polygons around the region of interest to project the alpha peak, which is result of signals coming from both detectors (the thin ΔE and the thick E) and then we fitted the projection. The FWHM's obtained for the complete 8 hours runs in periods of 2 hours are presented in Table 2. The results obtained showed no more changes than purely statistical ones, proving there was no alteration produced on the detectors caused by the jet action.

Table 2: Results of the FWHM fits for the projection of alphas peak from the ΔE-E telescope without jet and with jet for two hours periods in 8 hours runs.

runs of 2 h	FWHM (± 0.001 MeV) Jet OFF	FWHM (± 0.001 MeV) Jet ON
01 (0 to 2 h)	0.382	0.375
02 (2 to 4 h)	0.387	0.376
03 (4 to 6 h)	0.391	0.378
04 (6 to 8 h)	0.386	0.378

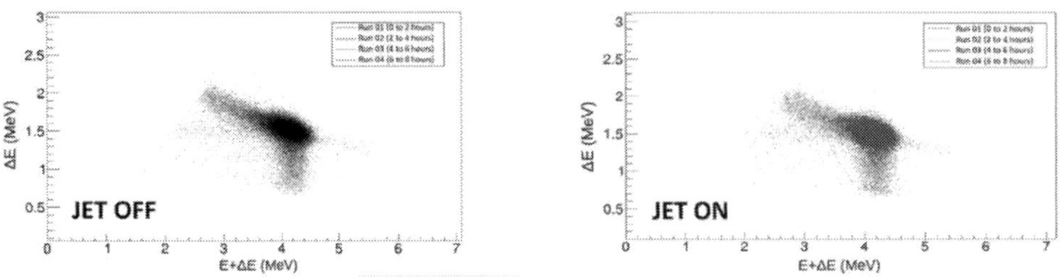

(a) (b)

Figure 5: ΔE-E_{tot} superposed spectra taken for 2 hours periods from 8 hours runs (a) without jet and (b) with jet working. There are not observable changes between both cases.

Analogously, the FWHM values for the P (front) side and N (back) side of the DSSSD were obtained. In Table 3 two examples for front and back strips are presented. From data table it can be seen that for strip 0P and for strip 2N of the DSSSD detector the FWHM value shows purely statistical variations for each 2 hours run, during the 8 hours of data collection with jet. Additionally, the Fig. 6 shows that in the four hitmaps the area with the highest concentration of incident particles is located in the lower right area, where strips 8-15P and 0-10N are. This indicates the position of the radioactive source, which remains constant in size and position. All of the above indicates that the DSSSD was not degraded or damaged in the presence of the jet, as neither happened in the telescope case.

Table 3: Results for two strips (front and back) of DSSSD FWHM measurements for jet data collection.

Strip	run	FWHM (± 0.001 MeV)
0P (front)	01 (0 to 2 h)	0.384
0P (front)	02 (2 to 4 h)	0.393
0P (front)	03 (4 to 6 h)	0.396
0P (front)	04 (6 to 8 h)	0.400
2N (back)	01 (0 to 2 h)	0.457
2N (back)	02 (2 to 4 h)	0.465
2N (back)	03 (4 to 6 h)	0.463
2N (back)	04 (6 to 8 h)	0.511

4 Conclusions

In the present work, the study on the evaluation of the behaviour of three different types of silicon semiconductor detectors (two of them very thin) when interacting with a supersonic jet target was carried out. The results presented here indicate that thin silicon detectors can be safely used in the presence of the SUGAR jet, without being damaged, for a real experiment. These results provide confidence for the mounting of sophisticate silicon arrays inside chambers with similar conditions as the SUGAR one.

The use of thin detectors is very important for pure jet targets experiments, considering the importance of particle discrimination in measurements at very low energies. At the same time, the use of thin DSSSD, besides providing us with a high granularity discrimination, which can give us an important increase of the solid angle, something as well very important

 https://doi.org/10.1051/epjconf/202533303003

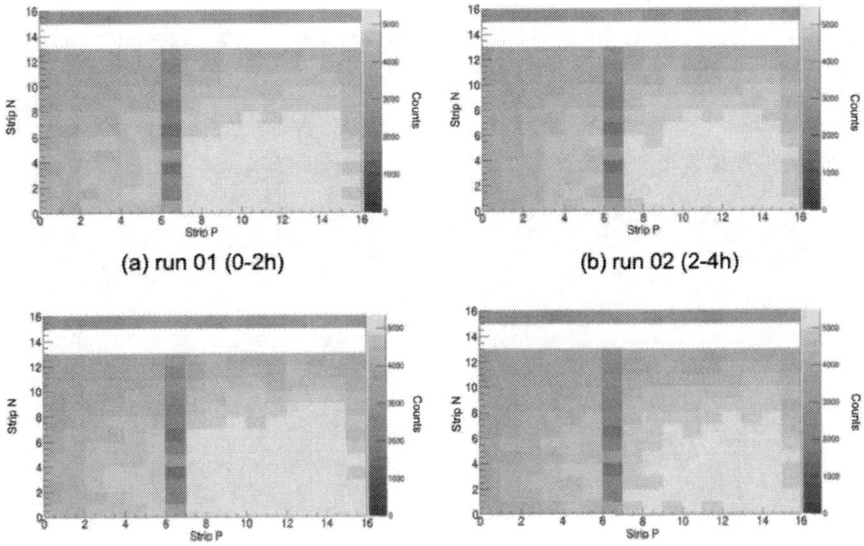

(a) run 01 (0-2h) (b) run 02 (2-4h)

(c) run 03 (4-6h) (d) run 04 (6-8h)

Figure 6: HitMap (N vs. P strips) for each 2 hours during 8 hours of data collection for DSSSD with jet. The enormous similarity between them demonstrate de absence of damage coming from the jet presence.

for low energy measurements. Particularly for the SUGAR case, the present findings increase its versatility to be used in different facilities around the world, where a target with its characteristics might be mounted.

This work was possible thanks to the financing of the Mexican projects DGAPA-UNAM PAPIIT IG101423, PAPIIT TA100524, Apoyos LNC-2023-58 CONAHCyT, CONAHCyT CBF2023-2024-862 and the Spanish MICIU PID2023-147569NB-C21 project.

References

[1] Favela, J., et. al Phys. Rev. S. T. 18, 123502. **1-4** (2015).

[2] COMSOL software https://www.comsol.com/

[3] Valcovic, V. *Radioactivity in the Environment.* Second Edition. Elsevier BV. **145** (2019).

[4] Choppin, G. et al. *Radiochemistry & Nuclear Chemistry.* 4th Edition. Elsevier Inc. (2013).

[5] Gotoh, S. & Takagi, Z. J. of Nuc. Sci. and Tec., 1, 8, **33** (1964).

[6] Posadillo, M. *Desarrollo de un prototipo con DSSSDs para realizar imagen médica con protones.* Facultad de Ciencias, Universidad de Sevilla. **9** (2019).

[7] Mesytec. (2025) *MPR module.* https://www.mesytec.com

[8] GSI Helmholtzzentrum für Schwerionenforschung GmbH. (n.d.). *FEBEX-3 data acquisition system.* https://www.gsi.de/en/

[9] Van, I. *Camac Multicrate Systems.* Astronomical Society of the Pacific, Volume 84. NAS Astrophysics Data System. **212** (1971).

[10] ORTEC *Model 512 amplifier.* https://www.ortec-online.com/products/electronic-instruments/amplifiers

[11] SRIM official web page. http://www.srim.org.

[12] William, L. *Techniques for Nuclear and Particle Physics Experiments: a how-to approach.* Springer-Verlag Berlin Heidelberg. **223** (1994).

EPJ Web of Conferences **333**, 03004 (2025) https://doi.org/10.1051/epjconf/202533303004
XLVI Symposium on Nuclear Physics 2025

Cryogenic Systems for the TUCAN EDM Experiment

Jeffery W. Martin[1,*,**] *B.* Algohi[2], *D.* Anthony[3], *L.* Barrón-Palos[4], *M.* Bradley[5], *A.* Brossard[3], *T.* Bui[2], *J.* Chak[3], *C.* Davis[3], *R.* de Vries[1], *K.* Drury[3], *D.* Fujimoto[3], *R.* Fujitani[6], *M.* Gericke[2], *P.* Giampa[3], *R.* Golub[7], *T.* Hepworth[1], *T.* Higuchi[8,9], *G.* Ichikawa[10], *S.* Imajo[9], *A.* Jaison[2], *B.* Jamieson[1], *M.* Katotoka[1], *S.* Kawasaki[10], *M.* Kitaguchi[11], *W.* Klassen[12], *E.* Korkmaz[13], *E.* Korobkina[7], *M.* Lavvaf[2], *T.* Lindner[3,1], *N.* Lo[3], *S.* Longo[2], *K.* Madison[12], *Y.* Makida[10], *J.* Malcolm[3], *J.* Mammei[2], *R.* Mammei[1], *C.* Marshall[3], *M.* McCrea[1], *E.* Miller[12], *M.* Miller[14], *K.* Mishima[11], *T.* Mohammadi[2], *T.* Momose[12], *T.* Okamura[10], *H.J.* Ong[9], *R.* Patni[3], *R.* Picker[3,15], *W.D.* Ramsay[3], *W.* Rathnakela[2], *J.* Sato[11], *W.* Schreyer[3,16], *T.* Shima[9], *H.* Shimizu[11], *S.* Sidhu[3], *S.* Stargardter[2], *P.* Switzer[1], *I.* Tanihata[9], *S.* Vanbergen[12], *W.T.H.* van Oers[2,3], *Y.* Watanabe[10], *A.* Zahra[2], and *M.* Zhao[3]

[1]The University of Winnipeg, Winnipeg, MB, Canada
[2]University of Manitoba, Winnipeg, MB, Canada
[3]TRIUMF, Vancouver, BC, Canada
[4]Universidad Nacional Autónoma de México, Mexico City, Mexico
[5]University of Saskatchewan, Saskatoon, SK, Canada
[6]Department of Nuclear Engineering, Kyoto University, Kyoto, Japan
[7]North Carolina State University, Raleigh, NC, USA
[8]Institute for Integrated Radiation and Nuclear Science (KURNS), Kyoto University, Osaka, Japan
[9]Research Center for Nuclear Physics (RCNP), Osaka University, Osaka, Japan
[10]High Energy Accelerator Research Organization (KEK), Tsukuba, Ibaraki, Japan
[11]Nagoya University, Aichi, Japan
[12]The University of British Columbia, Vancouver, BC, Canada
[13]The University of Northern BC, Prince George, BC, Canada
[14]McGill University, Montreal, QC, Canada
[15]Simon Fraser University, Burnaby, BC, Canada
[16]Oak Ridge National Laboratory, Knoxville, TN, USA

Abstract.
The TUCAN (TRIUMF UltraCold Advanced Neutron) Collaboration is completing a new ultracold neutron (UCN) source. The UCN source will deliver UCNs to a neutron electric dipole moment (EDM) experiment. The EDM experiment is projected to be capable of an uncertainty of 1×10^{-27} *e*cm, competitive with other planned projects, and a factor of ten more precise than the present world's best. The TUCAN source is based on a UCN production volume of superfluid helium (He-II), held at 1 K, and coupled to a proton-driven spallation target. The production rate in the source is expected to be in excess of 10^7 UCN/s; since UCN losses can be small in superfluid helium, this should allow us to build up a large number of UCNs. The spallation-driven superfluid helium technology is the principal aspect making the TUCAN project unique. The superfluid production volume was recently cooled, for the first time, and successfully filled with superfluid helium. The design principles of the UCN

*e-mail: j.martin@uwinnipeg.ca
**on behalf of the TUCAN Collaboration

source are described, along with some of the challenging cryogenic milestones that were recently passed.

1 Scientific motivation: the neutron electric dipole moment

The neutron electric dipole moment (nEDM) is an experimental observable of high importance in fundamental physics because it violates time-reversal symmetry and therefore CP (charge-parity) symmetry [1–3], the symmetry relating the interactions of particles to those of their antiparticle counterparts. To date, all experiments have found the nEDM to be compatible with zero. Improving the experimental precision places tighter constraints on new sources of CP violation beyond the Standard Model. Conversely, if a small but non-zero nEDM were discovered, it would herald a discovery of new physics. Even if ascribed to the CP-violating $\bar{\theta}$ parameter of the strong sector, the mystery of a small but non-zero $\bar{\theta}$ would create a new problem for the Standard Model.

A recent measurement performed using ultracold neutrons (UCNs) at the Paul Scherrer Institute (PSI) determined an upper bound on the nEDM, $|d_n| < 1.8 \times 10^{-26}$ $e$cm (90% C.L.) [4]. In addition to setting a new world record in precision, this work is noteworthy in that it is the first nEDM measurement conducted using a superthermal UCN source, a strategy pursued by our project and a host of new UCN sources that are expected to revolutionize the field.

Recent theoretical work addressing the physics impact of an even more precise measurement of the nEDM has focused on three general (and overlapping) themes: (1) new sources of CP violation beyond the Standard Model [5, 6], (2) baryogenesis scenarios, especially new physics contributions to electroweak baryogenesis inspired scenarios [7, 8] and (3) the strong CP problem, related to searches for axions [9–12]. Because of these connections, better measurements of the nEDM are of vital importance in particle physics and early universe cosmology.

Next generation UCN EDM experiments are in preparation at a variety of sites, worldwide, and are aiming to improve the result by an order of magnitude or more. Experiments are planned at Institut Laue-Langevin (ILL, Grenoble, France) [13], PSI [14], and Los Alamos National Laboratory (LANL, Los Alamos, NM, USA) [15, 16], and include our effort at TRIUMF [17]. Our goal of $\sigma(d_n) < 10^{-27}$ $e$cm is competitive with these efforts.

The TRIUMF UltraCold Neutron (TUCAN) Collaboration represents a collaboration of physicists from Canada, Japan, México, and the United States. Our experiment is conducted at TRIUMF, Canada's Particle Accelerator Centre, located in Vancouver, British Columbia. It uses a high-energy, high-intensity proton beam from the TRIUMF cyclotron impinging on a water-cooled tungsten target to initiate spallation reactions which liberate neutrons.

The basis of the TUCAN approach involves a spallation-driven, superfluid ^{4}He UCN source connected to a room-temperature EDM experiment. The key component making our project unique is our UCN source which, after the completion of our upgrade, is expected to become the world's best. We envision achieving UCN counting rates over 100 times larger than the last experiment at PSI, similar to or surpassing the plans of other experiments, and enabling the next breakthrough for this field.

2 Experiment overview

Figure 1 shows a schematic diagram of the planned apparatus. At the present time, all components of the UCN source have been completed, except for the liquid deuterium cryostat.

 https://doi.org/10.1051/epjconf/202533303004

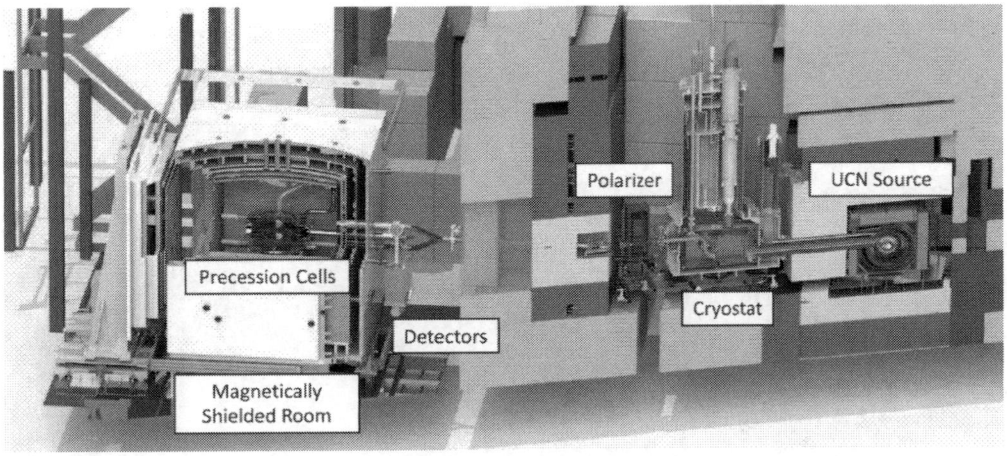

Figure 1. UCN source and EDM spectrometer for the TUCAN project. UCN exiting the source are polarized by a superconducting magnet and pass through UCN guides to reach the EDM experiment located within the MSR. UCN spins precess in dual EDM measurement cells with a holding field provided by a B_0 coil and electric field provided by a central HV electrode. UCN spin analyzers sense the neutron spins at the end of each cycle.

The UCN source itself will be described in further detail in Section 3. All components of the nEDM experiment have been prototyped and final versions are being built. The magnetically shielded room (MSR) needed for the nEDM experiment has been completed and has been shown to meet the specifications needed for a 10^{-27} $e$cm measurement.

An overhead view of the facility is shown in Fig. 2. The photograph was taken in April 2024, just before the UCN source was covered in shielding blocks. The facility is now complete to the stage that the UCN source has been operated in a month-long cryogenic test run, and the magnetically shielded room (MSR) is routinely in use for magnetometer and coil testing, in preparation for UCN runs planned for late 2025.

3 Ultracold neutron source principles and design

Our UCN source is based on previous work reported in Refs. [18, 19]. The prototype vertical UCN source developed in Japan was moved to Canada and installed in the Meson Hall at TRIUMF in 2017. We developed and constructed a new spallation target and proton beamline at TRIUMF for operation up to 40 μA (in preparation for an upgraded UCN source) [20]. The beamline features a fast kicker system which allows us to run simultaneously with other Meson Hall users [21]. We did experiments with the vertical source on UCN production [22], transport and storage [23], and polarization and detection [24]. In 2020-21, the vertical source was decommissioned in preparation for the upgrade.

The new UCN source upgrade features a 27 L UCN production volume (the He-II volume in Fig. 3) which is significantly larger than the 8 L volume used in the vertical source. The vertical source could reliably handle 300 mW of heat load to the He-II whereas the horizontal source is optimized to handle 10 W [25, 26]. A new, larger capacity helium pumping system enables the additional cooling power. Additionally, a new large-area ^{3}He-^{4}He heat exchanger (Fig. 3) was built to be compatible with both UCN transport and heat transfer requirements, resolving a severe limitation of the vertical source [24].

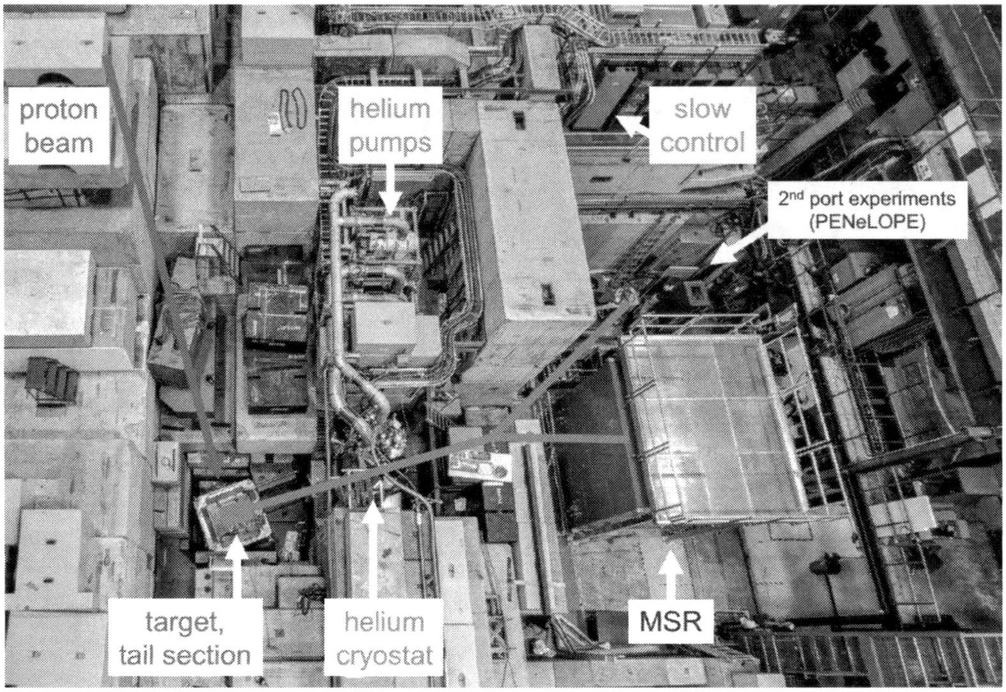

Figure 2. Overhead view of the UCN source facility (April 2024). Lines display the underlying proton beam path (red) and sketch the existing and planned UCN guide paths (blue).

We completed detailed estimates for UCN production and extraction [23, 27, 28] based on a Monte Carlo N-Particle (MCNP) model of the source, a model of UCN production based on Ref. [29], and UCN transport simulations based on Ref. [30] including losses within the He-II and in transport to the EDM experiment. The optimization indicates that when driven by a 40 μA proton beam, the source will produce 1.4×10^7 UCN/s, with beam heating of 8.1 W to the He-II at 1.1 K. This is more than two orders of magnitude larger than the UCN production rate of the vertical source. A total of 1.38×10^7 UCN would be loaded into the EDM measurement cells prior to initiating the Ramsey cycle. Using reasonable values for lifetimes and spin-coherence times of the UCN, this corresponds to a statistical determination of the nEDM of $\sigma(d_n) = 3 \times 10^{-25}$ $e$cm per cycle. Using conservative assumptions for the running time available per day, a statistical determination of $\sigma(d_n) = 10^{-27}$ $e$cm would be achieved within 280 days of running [27].

4 Recent cryogenic performance results

In experiments up to November 2024, the source was cooled and filled with superfluid ^{4}He. During this time, the LD_2 cryostat was not in place. The cryogenic performance of the source was excellent. A base temperature of 0.8 K was achieved with an acceptable resting heat load. No evidence of a superleak was seen. No clogs of either the ^{3}He or natural-abundance helium systems were experienced in >20 days of operation. The temperatures in the tail section were consistent with the superfluid helium being maintained at ~1 K. The beam heat load was measured (see Fig. 4) and was consistent with expectations based on MCNP simulations within 10%. There was some evidence of a rise of temperature sensors closer to the UCN

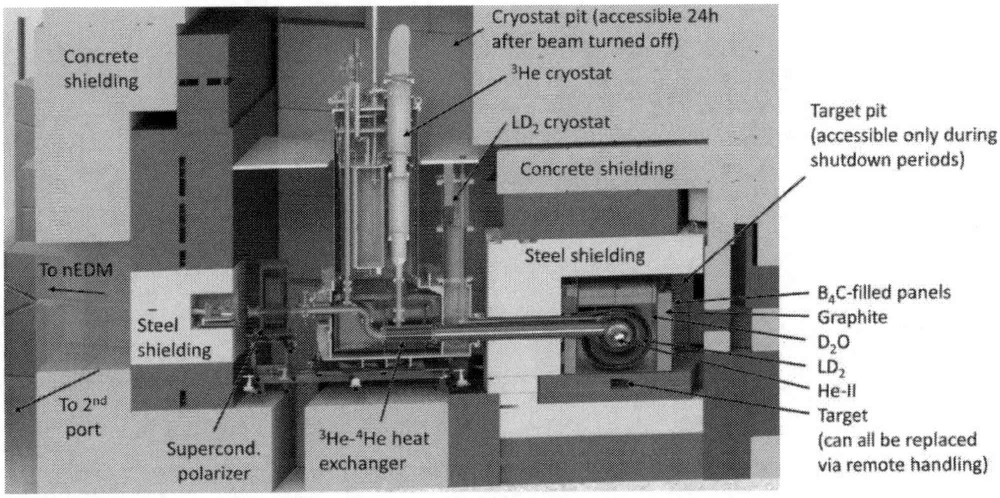

Figure 3. The recently completed UCN source. Neutrons are liberated by proton-induced spallation at 480 MeV and 40 μA in a target located beneath the He-II, LD_2, and D_2O volumes. Neutrons are reflected and moderated in surrounding materials then enter superfluid ^{4}He (He-II) where they are down-scattered to become UCNs. UCNs created in the He-II are transported out through the heat exchanger passing through the superconducting polarizer magnet to the nEDM experiment.

production volume under the highest beam heat load, consistent with our expectations based on heat conduction in turbulent He-II (the Gorter-Mellink regime).

The heat exchanger and refrigerator were capable of maintaining sufficiently low ^{4}He temperatures, in heater tests mimicking our highest projected beam heat load of 10 W. An example of this type of test is presented in Fig. 5, where the ^{3}He pot temperature is reduced to 0.9 K when the Joule-Thomson needle valve is opened sufficiently. During these tests, the ^{3}He flow rate was measured and, as in the beam heating tests, were found to be consistent with expectation.

In runs where beam was present, we attempted to sense UCNs in a detector connected directly to the UCN source (in place of the superconducting polarizer in Fig. 3). No conclusive evidence of UCN detection was seen above the larger background in this region, despite $\sim 10^4$ UCNs/μA being expected after saturation of the UCN density (a 60 s irradiation time). We strongly suspect this was caused by contamination of the ^{4}He production volume with either air or water frozen on the inner surface. As we were condensing ^{4}He, we experienced clogging of the condensation route in the cryostat, which necessitated filling inefficiently through the recovery line. This means that contaminants could freeze on the coldest parts of our cryostat, in the production volume.

Prior to our next runs, planned for June 2025, we plan to purify the ^{4}He using both our UCN source cryostat, and a new purifier system from Japan, thus preventing future clogs. This will allow us to fill the cryostat in the proper way that it was designed to operate. First detection of UCN produced by the new source will be the key milestone that we aim to achieve in the next runs. This is a very exciting time to be operating the facility, as we turn on the UCN source for the first time and begin to verify its capabilities.

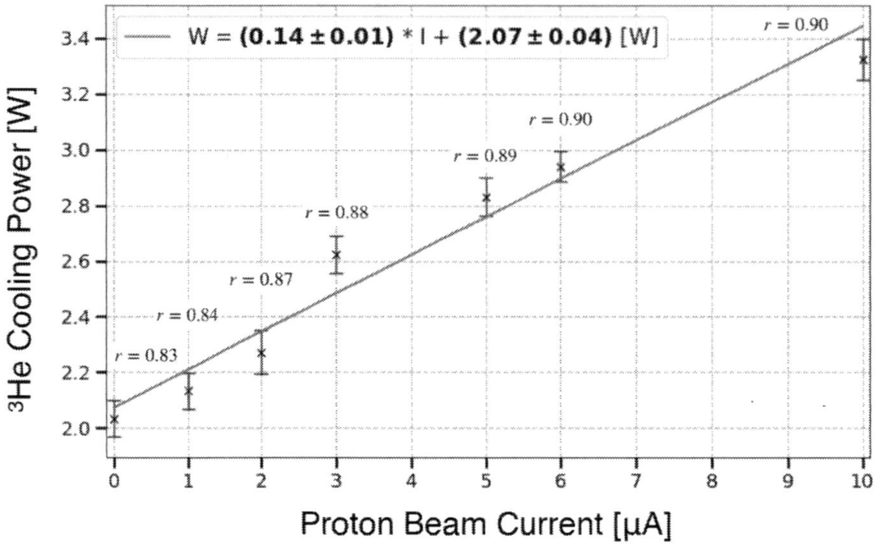

Figure 4. Measurement of heat removed by ^{3}He pumping as a function of beam current delivered to the spallation target, at a ^{3}He temperature of 0.9 K; r is the fraction of liquid ^{3}He remaining after Joule-Thomson expansion.

5 Future Plans and Conclusion

The TUCAN project has made incredible progress in the past year, with the cryogenic commissioning of the superfluid helium UCN source now complete. The next major installation for the project is the liquid deuterium (LD_2) cryostat. This is scheduled for spring 2025, and is needed to boost UCN production by a factor of 30, which will make the UCN source truly world-class. The collaboration aims to complete the milestones of first UCN detection, commissioning of the LD_2 cryostat, and delivery of UCN into the magnetically shielded room, by the end of the calendar year in 2025.

In 2026, the TRIUMF laboratory will undergo a year-long shutdown in support of the ARIEL project. In 2027, the schedule calls for commissioning of the TUCAN EDM experiment and preparation for data taking.

Simultaneously with these efforts, we are applying for funding for an upgrade to our helium liquefier facility, from the Canada Foundation for Innovation and funding sources in Japan. Included in the upgrade are further improvements to the UCN source for UCN production and delivery, improvements to the nEDM experiment to implement the two-measurement-cell system shown in Fig. 1, and funding for further research into dual-species Xe-Hg comagnetometry. These developments would be completed in 2028 and beyond.

Acknowledgments

We gratefully acknowledge the support of the Canada Foundation for Innovation; the Canada Research Chairs program; the Natural Sciences and Engineering Research Council of Canada (NSERC) SAPPJ-2016-00024, SAPPJ-2019-00031, and SAPPJ-2023-00029; JSPS KAKENHI (Grant Nos. 18H05230, 19K23442, 20KK0069, 20K14487, and 22H01236); JSPS

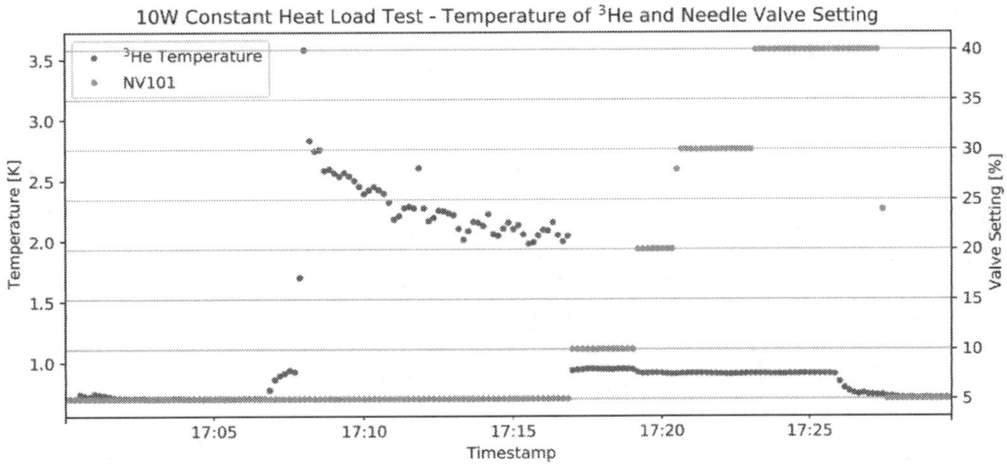

Figure 5. Temperature of the ^{3}He in the ^{3}He-^{4}He copper heat exchanger, as a function of time during the application of 10 W of heat. Blue points and left axis: ^{3}He temperature. Orange points and right axis: Joule-Thomson needle valve setting. The ^{3}He temperature rapidly reduces to 0.9 K when the needle valve is opened sufficiently. Before and after the application of the 10 W of heat, the ^{3}He temperature is 0.7 K.

Bilateral Program (Grant No. JSPSBP120239940); JST FOREST Program (Grant No. JP-MJFR2237); International Joint Research Promotion Program of Osaka University; RCNP COREnet; the Yamada Science Foundation; the Murata Science Foundation; the Grant for Overseas Research by the Division of Graduate Studies (DoGS) of Kyoto University; and the Universidad Nacional Autónoma de México - DGAPA program PASPA and grant PAPIIT AG102023.

References

[1] M. Pospelov and A. Ritz, "Electric dipole moments as probes of new physics," Ann. Phys. **318** 119-169 (2005). https://doi.org/10.1016/j.aop.2005.04.002

[2] J. Engel, M.J. Ramsey-Musolf, U. van Kolck, "Electric dipole moments of nuclei, nucleons, and atoms," Prog. Part. Nucl. Phys. **71**, 21 (2013). https://doi.org/10.1016/j.ppnp.2013.03.003

[3] T. E. Chupp, *et al.*, "Electric dipole moments of atoms, molecules, nuclei, and particles," Rev. Mod. Phys. **91**, 015001 (2019). https://doi.org/10.1103/RevModPhys.91.015001

[4] C. Abel, *et al.*, "Measurement of the Permanent Electric Dipole Moment of the Neutron," Phys. Rev. Lett. 124, 081803 (2020). https://doi.org/10.1103/PhysRevLett.124.081803

[5] V. Cirigliano, *et al.*, "CP Violation in Higgs-Gauge Interactions: From Tabletop Experiments to the LHC," Phys. Rev. Lett. **123**, 051801 (2019). https://doi.org/10.1103/PhysRevLett.123.051801

[6] A. Crivellin and F. Saturnino, "Correlating tauonic B decays with the neutron electric dipole moment via a scalar leptoquark," Phys. Rev. D **100**, 115014 (2019). https://doi.org/10.1103/PhysRevD.100.115014

[7] N. F. Bell, *et al.*, "Electric dipole moments from postsphaleron baryogenesis," Phys. Rev. D **99**, 015034 (2019). https://doi.org/10.1103/PhysRevD.99.015034

[8] W.-S. Hou, G. Kumar, and S. Teunissen, "Discovery prospects for electron and neutron electric dipole moments in the general two Higgs doublet model" Phys. Rev. D **109**, L011703 (2024). https://doi.org/10.1103/PhysRevD.109.L011703

[9] M. Carena, *et al.*, "ν solution to the strong CP problem," Phys. Rev. D **100**, 094018 (2019). https://doi.org/10.1103/PhysRevD.100.094018

[10] Y. Mimura, R. N. Mohapatra, and M. Severson, "Grand unified parity solution to the strong CP problem," Phys. Rev. D 99, 115025 (2019). https://doi.org/10.1103/PhysRevD.99.115025

[11] R. Ferro-Hernandez, S. Morisi, and E. Peinado, "Axionless strong CP problem solution: the spontaneous CP violation case," arXiv:2407.18161 (2024). https://doi.org/10.48550/arXiv.2407.18161.

[12] C. Abel *et al.*, "Search for Axionlike Dark Matter through Nuclear Spin Precession in Electric and Magnetic Fields," Phys. Rev. X **7**, 041034 (2017). https://doi.org/10.1103/PhysRevX.7.041034.

[13] D. Wurm *et al.*, "The PanEDM Neutron Electric Dipole Moment Experiment," EPJ Web Conf. **219**, 02006 (2019). https://doi.org/10.1051/epjconf/201921902006

[14] N.J. Ayres *et al.*, "The design of the n2EDM experiment," Eur. Phys. J. C **81**, 512 (2021). https://doi.org/10.1140/epjc/s10052-021-09298-z

[15] R. Alarcon, these proceedings.

[16] D. K.-T. Wong, *et al.*, "Characterization of the new Ultracold Neutron beamline at the LANL UCN facility," Nucl. Instrum. Meth. A **1050**, 168105 (2023). https://doi.org/10.1016/j.nima.2023.168105

[17] J. W. Martin, *et al.*, "The TRIUMF UltraCold Advanced Neutron Source," Nucl. Phys. News **31**, 19 (2021). https://doi.org/10.1080/10619127.2021.1881367

[18] Y. Masuda *et al.*, "Spallation Ultracold-Neutron Production in Superfluid Helium," Phys. Rev. Lett. **89**, 284801 (2002). https://doi.org/10.1103/PhysRevLett.89.284801

[19] Y. Masuda *et al.*, "Spallation Ultracold Neutron Source of Superfluid Helium below 1 K," Phys. Rev. Lett. **108**, 134801 (2012). https://doi.org/10.1103/PhysRevLett.108.134801

[20] S. Ahmed *et al.* (TUCAN Collaboration), "A beamline for fundamental neutron physics at TRIUMF," Nucl. Instrum. Meth. A **927**, 101 (2019). https://doi.org/10.1016/j.nima.2019.01.074

[21] S. Ahmed *et al.* (TUCAN Collaboration), "Fast-switching magnet serving a spallation-driven ultracold neutron source," Phys. Rev. Accel. Beams **22**, 102401 (2019). https://doi.org/10.1103/PhysRevAccelBeams.22.102401

[22] S. Ahmed *et al.* (TUCAN Collaboration), "First ultracold neutrons produced at TRIUMF," Phys. Rev. C 99, 025503 (2019). https://doi.org/10.1103/PhysRevC.99.025503

[23] W. Schreyer *et al.*, "Optimizing neutron moderators for a spallation-driven ultracold-neutron source at TRIUMF," Nucl. Instrum Meth. A **959** 163525 (2020). https://doi.org/10.1016/j.nima.2020.163525

[24] S. Hansen-Romu, "Modeling He-II cryostat performance and characterization of spin manipulation components for a neutron electric dipole moment experiment at TRIUMF," PhD thesis, U. Manitoba (2023). https://mspace.lib.umanitoba.ca/items/f834c696-740c-4bcb-ac48-ea7222b596a4

[25] S. Kawasaki et al., "Development of a Helium-3 Cryostat for a Ultra-Cold Neutron Source," IOP Conf. Ser.: Mater. Sci. Eng. 755, 012140 (2020). https://doi.org/10.1088/1757-899X/755/1/012140

[26] T. Okamura et al., "Thermo-fluid analyses for UCN cryogenic system," IOP Conf. Ser.: Mater. Sci. Eng. 755, 012141 (2020). https://doi.org/10.1088/1757-899X/755/1/012141

[27] S. Sidhu, *et al.*, "Estimated performance of the TRIUMF ultracold neutron source and electric dipole moment apparatus," EPJ Web of Conf. **282**, 01015 (2023). https://doi.org/10.1051/epjconf/202328201015

[28] S. Sidhu, "Improving the statistical sensitivity reach of the TUCAN neutron electric dipole moment experiment," PhD thesis, Simon Fraser U. (2024). https://summit.sfu.ca/item/36485

[29] E. Korobkina, R. Golub, B.W. Wehring, and A.R. Young, "Production of UCN by downscattering in superfluid He-4," Phys. Lett. A **301** 462 (2002). https://doi.org/10.1016/S0375-9601(02)01052-6

[30] W. Schreyer, *et al.*, "PENTrack—a simulation tool for ultracold neutrons, protons, and electrons in complex electromagnetic fields and geometries," Nucl. Instrum. Meth. A **858**, 123 (2017). https://doi.org/10.1016/j.nima.2017.03.036

Improvement of a plastic scintillator with different coatings for gamma-ray detection

J. Méndez-García[1], *H.* Alva-Sánchez[1], *L.C.* Alvarez-Núñez[2], *E.* Chavéz[1], *A.* Farah-Simon[2], *A.* García-Flores[1], *A.* Huerta[1], *O.* López-López[1], *D. J.* Marín-Lámbarri[1,*], *T.* Murrieta-Rodriguez[1], and *C.* Valencia[1]

[1]Instituto de Física, Universidad Nacional Autónoma de México, Apartado Postal 20-364, 01000 Ciudad de México, México.
[2]Instituto de Astronomía, Universidad Nacional Autónoma de México, Ciudad de México, México.

Abstract. Identical cubic plastic scintillator detectors were tested using different coating materials in order to investigate their response function and efficiency, additionally two light guides were also coated evaluating their performance in order to find the best combination for a detection array. The array is composed by a plastic scintillator coupled to a light guide and a photomultiplier tube (PMT), the output signal is processed by a Data Aquisition System (DAQ). Operational voltage for the PMT was found and detectors were tested measuring ^{137}Cs certified gamma-ray point source. Spectra were analyzed and optimal coating for light guide and detector combination was determined based on the array's efficiency.

1 Introduction

Gamma-ray spectroscopy is a highly effective tool employed to measure nuclear reactions for the characterization of nuclei [1]. Gamma-rays can be detected using different type of detectors, including high-purity germanium detectors, which offer good energy resolution, as well as organic and inorganic scintillators coupled to photomultiplier tubes (PMTs) or silicon photomultipliers (SiPMs) [2–7]. While the latter provides the lower energy resolution, they exhibit higher detection efficiency, with both types converting detected photons into electric current. In this way the plastic scintillators results as a good option, being less expensive than the other detector types, with advantages like its high light output and a fast signal response, short decay time, good hardness radiation and their ability to be shaped according to the experimental requirements[8–12]. Plastic scintillators are also an option to measure charged particles with high detection efficiency, contrary to that for gamma-rays due its low atomic number composition [12], the latter being not a limitation to continue coping the challenge in nuclear physics experiments as the background radiation reduction, primary generated by cosmic-rays or radioactive materials present in the laboratory's vicinity. Shielding is typically implemented to attenuate this background radiation; passive shielding is often constructed from heavy metals like lead and copper in order to house the detector array, while active shielding employs scintillators to identify undesired signals and use them as veto [13–15].

*e-mail: dmarin@fisica.unam.mx

2 Experimental Set-up

Four identical cubes (42.75 x 38.00 x 50.00 mm) of BC-408 plastic scintillator were fabricated and coated with four different materials (PTFE or Teflon tape, black tape, white paint and silver) before being sequentially coupled to a PMT using two light guides (one coated with silver and the other with PTFE). The output signal from the PMT is an electrical pulse proportional to the number of photons interacting with the scintillator. The guides and scintillators are shown in Figure 1.

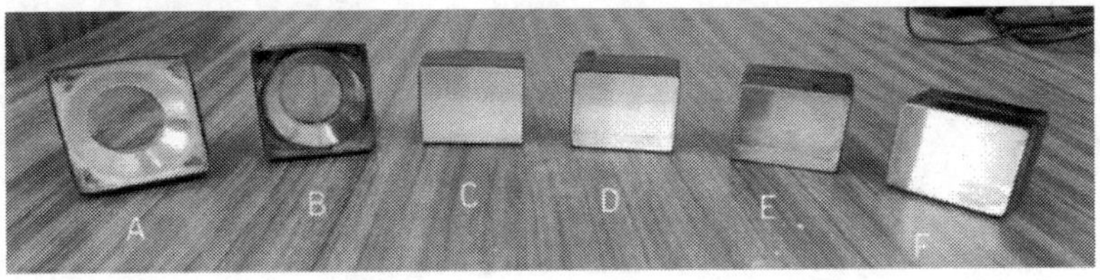

Figure 1. Optical guides and scintillator cubes: A) Light guide with PTFE tape. B) Light guide coated with silver. C) White-painted scintillator. D) PTFE covered scintillator. E) Silver and laquer coated scintillator. F) Silver coated scintillator.

To ensure that all the components are perfectly coupled, a wooden black box was fabricated and used as a case for the detector array, shielding it from light. The inside width of the box is the same as the width of the cubic scintillator in order to firmly couple the array and keep it on place, the height is larger than the cube dimension, enough to place the coupled components as shown in Figure 2.

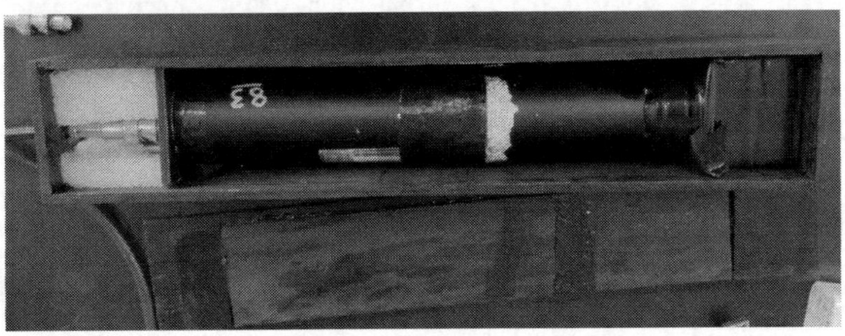

Figure 2. Detector array, from right to left, the plastic scintillator cube coupled to a light guide, coupled to a PMT.

3 Results and discussion

The measurements were performed with a certified ^{137}Cs gamma-ray source located in front of the box choosing first the no-coated scintillator and the white-tape coated light guide varying the source-box distance, then with all the different coated scintillators. Each measurement was performed in a 5 minutes period, with a total of 120 measurements, including a background radiation measurement for each combination. A spectrum was obtained with the silver-coated light guide and was switched through the different scintillators as shown in Figure 3. The integration results are shown in table 1.

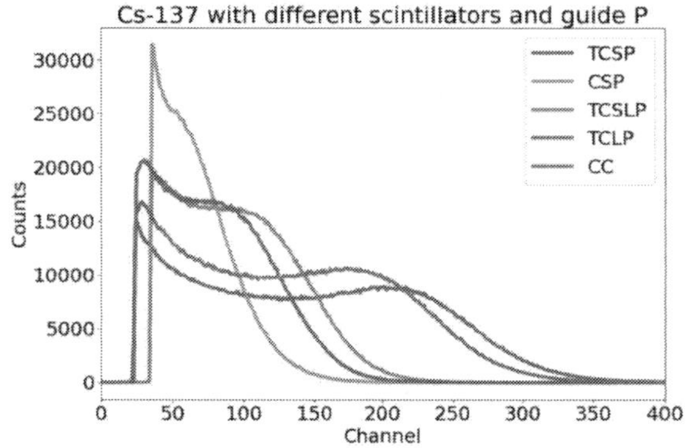

Figure 3. Spectra of ^{137}Cs using all the scintillators coupled each time to the silver-coated light guide. TCSP is the white-tape coated scintillator, CSP the no-coated one, TCSLP the no-protected silver cube, TCLP the protected silver one and CC the white-painted scintillator.

Table 1. Counts of photons for each coated scintillator in the array.

Coating	Counts ($\times 10^6$)
CSP	1.506
TCLP	1.920
TCSLP	1.980
TCSP	2.280
CC	2.478

The optical guides and scintillator cubes have different components so it present different behaviors in their transfer of light. After coating these materials a change in the response will be expected. In all cases using ^{137}Cs, the same behaviors were obtained in the spectra, where white coated guides yield a less energetic Compton edge and a higher number of counts as shown in Figure 4 where the white-painted scintillator was used and the light guides were switched.

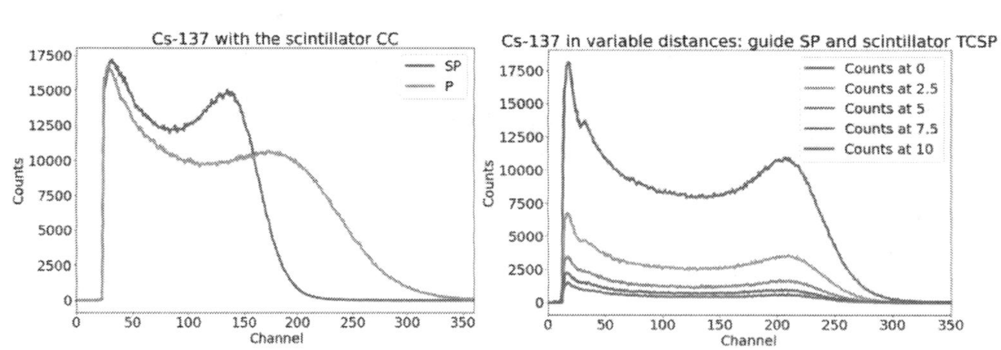

Figure 4. Left plot, spectra of ^{137}Cs using the white-painted scintillator switching through the two light guides. SP is the white-tape coated light guide and P is the silver-coated one. An integration under each curve shown that with the silver-coated the counts increase to 2.48 x 10^6 while with white-tape 2 x 10^6 was obtained. Right plot, Comparison of the spectra obtained varying the distance (at 0, 2.5, 5, 7.5 and 10 cm) between the scintillating cube with white tape and the ^{137}Cs source using the guide covered with white tape.

4 Conclusions

From Figure 3, we observed that for the light guide, the best coating material is silver due to its intrinsic geometry: the tilt angle in the light guide makes that the light rays are reflected to the PMT supported by the optical properties of the silver as a mirror that consist in reflecting all the photons in the same direction. On the other hand, for the scintillators, white-paint is the best option as shown in Table 1 because the geometry is completely different in this case: for a surface without a tilt angle the lose of photons is more significant with a silver coating but a white coating reflects light isotropically and as a result, the photons are transmitted towards the light guide.

Acknowledgments

This work has been partially funded by CONAHCYT CBF2023-2024-862, 2023-LNC-58 and UNAM PIIF-5-24, PAPIIT TA10052 and IN112023.

References

[1] Buchtela, K., Radiochemical Methods: Gamma-Ray Spectrometry. Encyclopedia of Analytical Science, 3rd edition, **9** (2019). https://doi.org/10.1016/B978-0-12-409547-2.11333-2.

[2] Knoll, G. F., Radiation Detection and Measurement (Arbor, Michigan: Wiley, 2010).

[3] Hirsch, P., Introduction to Radiation Detection (Wiley-Interscience, 2007).

[4] Swenney, T. and Byrnes H. Radiation Detection and Measurement (ohn Wiley Sons, 1990).

[5] Marín, D. J., Leal, B., Technical Report: Scintillator Detector for Real-Time Dosimetry of Intense Gamma Radiation Fields (Institute of Nuclear Sciences, UNAM) (2022).

[6] SI Photodiodes, Hamamatsu Photonics. (n.d.). https://www.hamamatsu.com/eu/en/product/optical-sensors/photodiodes/si-photodiodes.html

[7] Kharzheev, Y. N. Physics of Particles and Nuclei **50** 1 42-76 (2019) DOI:10.1134/S1063779619010027.

[8] https://physicsopenlab.org/2016/03/31/sipm-plastic-scintillator/

[9] Son, J. et. al. Improved 3D Printing Plastic Scintillator Fabrication. Journal of the Korean Physical Society **73**, 7 887-892 (2018). DOI:10.3938/jkps.73.887.

[10] L. Solutions, BC400, BC404, BC408, BC412, BC416. (s/f). Luxium Solutions. https://luxiumsolutions.com/radiation-detection-scintillators/plastic-scintillators/bc400-bc404-bc408-bc412-bc416

[11] Gobain. S., Scintillation materials and assemblies https://www.luxiumsolutions.com/sites/default/files/2021-09/Scintillation-Materials-and-Assemblies.pdf

[12] Zhang, Y. et. al. Radiation hardness study of BC408 plastic scintillator under 80 MeV proton beam irradiations. Nucl. Inst. and Met. in Phys. Res. B **548** 165247 (2024). https://doi.org/10.1016/j.nimb.2024.165247

[13] Beddar,A. Plastic scintillation dosimetry and its application to radiotherapy. Radiat. Meas. **41** S124–S133 (2006). https://doi.org/10.1016/j.radmeas.2007.01.002

[14] ATLAS Collab.,The ATLAS experiment at the CERN Large Hadron Collider. J. Instrum. **3** S08003 (2008).

[15] Liao, S. et. al. A comparative study of the radiation hardness of plastic scintillators for the upgrade of the Tile Calorimeter of the ATLAS detector. J. Phys. Conf. Ser. **645** 012021 (2015).

Status of the NUMEN Construction

D. Pierroutsakou[1,*], *L.* Acosta[2], *N.* Added[3], *C.* Agodi[4], *V.A.P.* Aguiar[3], *P.* Amador-Valenzuela[5], *A.* Anastasio[1], *N.* Auerbach[6], *L.H.* Avanzi[7], *Y.* Ayyad[8], *R.* Babu[9], *Y.* Babu[8], *R.* Bijker[10], *A.* Boiano[1], *I.* Boztosun[11], *S.* Brasolin[12], *G.A.* Brischetto[4], *M.P.* Bussa[12,13], *D.* Calvo[12], *F.* Cappuzzello[14,4], *D.* Carbone[4], *E.N.* Cardozo[15], *G.* Castro[4], *M.* Cavallaro[4], *K.* Challa[12], *E.R.* Chávez Lomelí[2], *E.F.* Chinaglia[7], *I.* Ciraldo[4], *M.* Colonna[4], *A.* Comite[16], *M.* Cortesi[17], *K.M.* Costa[7], *G.* D'Agostino[4], *H.* Dapo[18], *C.* De Benedictis[12,19], *G.* De Gregorio[20,1], *L.M.* Donaldson[21], *F.* Dumitrache[12], *A.* Huerta Hernandez[2], *D.* Gambacurta[4], *E.* M. Gandolfo[22], *H.* Garcia-Tecocoatzi[23], *A.* Gargano[1], *C.* Garofalo[14,4], *M.* Giovannini[16,24], *R.* Gleason[2], *A.* Hacisalihoglu[25], *C.* Ferraresi[12,26], *J.L.* Ferreira[15], *V.* Izzo[1], *S.* Koulouris[27], *Y.* Kucuk[11], *T.C.* Khumalo[21,28], *L.* La Fauci[4], *J.A.* Lay[29], *H.* Lenske[30], *R.* Linares[15], *C.* Lombardo[4], *J.M.* López González[8], *J.* Lubian[15], *D.J.* Marín-Lámbarri[2], *J.* Mas-Ruiz[10], *S.H.* Masunaga[7], *N.H.* Medina[3], *M.* Moralles[31], *L.* Neri[4], *R.* Neveling[21], *Thang* Nguyen[32], *J.R.B.* Oliveira[3], *A.* Pakou[33], *K.* Palli[27,33], *A.* Pandalone[1], *L.* Pandola[4], *R.* Panero[12], *M.* Paterna[12,19], *N.* Pietralla[34], *A.* Pitronaci[4], *L.* Pellegri[21,28], *R.* Persiani[35], *H.* Petrascu[36], *A.* Rovelli[4], *A.D.* Russo[4], *S.* Sandoval-Hipólito[2], *T.M.* Santarelli[7], *E.* Santopinto[24], *R.B.B.* Santos[7], *D.* Sartirana[12], *J.V.* Schervenin[37], *O.* Sgouros[4], *S.O.* Solakcı[11], *V.* Soukeras[4], *G.* Souliotis[27], *A.* Spatafora[4], *D.* Torresi[4], *B.* Urazbekov[38,39], *C.* Valencia[2], *I.* Valinotto[12], *A.* Vanzanella[1], *R.I.M.* Vsevolodovna[24], *A.* Yildirim[11], and *V.A.B.* Zagatto[15]

[1]Istituto Nazionale di Fisica Nucleare, Sezione di Napoli, Napoli, Italy
[2]Instituto de Física, Universidad Nacional Autónoma de México, México City, México
[3]Instituto de Fisica, Universidade de Sao Paulo, Sao Paulo, Brazil
[4]Istituto Nazionale di Fisica Nucleare, Laboratori Nazionali del Sud, Catania, Italy
[5]Instituto Nacional de Investigaciones Nucleares, México
[6]School of Physics and Astronomy, Tel Aviv University, Israel
[7]Centro Universitario FEI, Sao Bernardo do Campo, Brazil
[8]IGFAE, Universidade de Santiago de Compostela, Santiago de Compostela, Spain
[9]Istituto Nazionale di Fisica Nucleare, Sezione di Genova, Italy
[10]Instituto de Ciencias Nucleares, Universidad Nacional Autónoma de México, Mexico
[11]Department of Physics, Akdeniz University, Turkey
[12]Istituto Nazionale di Fisica Nucleare, Sezione di Torino, Torino, Italy
[13]Dipartimento di Fisica, Università di Torino, Italy
[14]Dipartimento di Fisica e Astronomia "Ettore Majorana", Università di Catania, Catania, Italy
[15]Instituto de Fisica, Universidade Federal Fluminense, Brazil
[16]Dipartimento di Chimica e Chimica Industriale, Università di Genova, Italy
[17]Michigan State University, East Lansing, MI, US
[18]Ankara University, Institute of Accelerator Technologies, Turkey
[19]DISAT, Politecnico di Torino, Torino, Italy
[20]University of Campania "Luigi Vanvitelli", Caserta, Italy
[21]iThemba Laboratory for Accelerator Based Sciences, Faure, Cape Town, South Africa
[22]GSI Helmholtzzentrum für Schwerionenforschung GmbH, Darmstadt, Germany
[23]Instituto Tecnológico y de Estudios Superiores de Monterrey, México

*dimitra.pierroutsakou@na.infn.it

[24]Istituto Nazionale di Fisica Nucleare, Sezione di Genova, Genova, Italy
[25]Recep Tayyip Erdogan University, Department of Physics, Turkey
[26]DIMEAS, Politecnico di Torino, Torino, Italy
[27]Department of Chemistry, National and Kapodistrian University of Athens and HINP, Greece
[28]School of Physics, University of the Witwatersrand, Johannesburg, South Africa
[29]Departamento de FAMN, University of Seville, Spain
[30]Department of Physics, University of Giessen, Germany
[31]Instituto de Pesquisas Energeticas e Nucleares, Sao Paulo, Brazil
[32]Florida State University, Florida, US
[33]Department of Physics and HINP, University of Ioannina, Greece
[34]Institut fur Kernphysik, Technische Universitat Darmstadt, Germany
[35]Istituto Nazionale di Fisica Nucleare, Sezione di Catania, Catania, Italy
[36]Department of Nuclear Physics, Horia Hubulei Nat. Inst. Phys. and Nucl. Eng., IFIN-HH, Romania
[37]Instituto de Fisica, Universidade de Sao Paulo, Brazil
[38]Department of Nuclear Physics and Nanotechnology, L. N. Gumilyov Eurasian National University, Astana, Kazakhstan
[39]Institute of Nuclear Physics, Laboratory of Low Energy Reactions, Almaty, Kazakhstan

Abstract.
The NUMEN (NUclear Matrix Elements for Neutrinoless double beta decay) project at INFN-Laboratori Nazionali del Sud aims to extract quantitative information on the Nuclear Matrix Elements relevant to neutrinoless double beta decay, a key process for determining whether the neutrino is a Majorana or Dirac particle and for establishing its effective mass. This objective will be pursued by measuring the cross sections of double charge exchange reactions induced by intense heavy-ion beams on selected isotopes that are candidates for neutrinoless double beta decay. The need to measure extremely low cross sections with high statistical significance has driven a major upgrade of the entire INFN-LNS infrastructure, enabling the production of heavy-ion beams with intensities up to 10^{13} pps at the position of the experimental target. Additionally, significant enhancements are being made to the focal plane detectors of the existing MAGNEX spectrometer. A new target system and advanced detectors are under development to track and identify heavy ions at an expected rate of 5×10^6 Hz at full beam intensity, while also allowing γ-ray detection. This contribution provides an overview of the current status of the NUMEN project, highlighting recent developments and characterization campaigns for nuclear targets and detector systems.

1 Introduction

Neutrinoless double beta decay ($0\nu\beta\beta$) is a hypothetical isobaric transition in which a parent nucleus (A,Z) transforms into a daughter nucleus (A,Z+2) with the emission of two electrons but no electron antineutrinos. Its observation, which is forbidden by the Standard Model, would provide direct evidence of lepton number violation, with profound implications for our understanding of the Universe's evolution and the matter-antimatter asymmetry. Moreover, it would determine the nature of the neutrino —whether it is a Dirac or Majorana particle— confirming the latter and costrain the effective neutrino mass, provided that the Nuclear Matrix Elements (NMEs) of the involved operators are accurately known. However, current calculations of NMEs, based on various nuclear structure models, exhibit significant uncertainties [1, 2].

Different approaches are being pursued to extract information on $0\nu\beta\beta$ NMEs through experimentally accessible observables. The NUMEN project aims to provide insights into

$0\nu\beta\beta$ NMEs by measuring the cross sections of heavy-ion Double Charge Exchange (DCE) reactions on isotopes that are candidates for $0\nu\beta\beta$ decay [3–5]. DCE reactions involve the exchange of two charge units between the projectile and the target while preserving the initial mass partition. These reactions can occur through sequential nucleon transfer mechanisms, which have been shown to make a negligible contribution [6], or through the exchange of two isovector charged mesons, either in an uncorrelated two-step single charge exchange process [7] —analogous to two-neutrino double beta decay ($2\nu\beta\beta$)— or in a correlated manner between target and projectile. The latter process, known as Majorana DCE, closely resembles $0\nu\beta\beta$ decay [8, 9], suggesting that their NMEs are related [3–5].

The NUMEN project plans to utilize oxygen, neon and carbon beams at energies ranging from 15 to 70 AMeV on several candidate isotopes of interest, including ^{48}Ca, ^{76}Ge, ^{82}Se, ^{100}Mo, ^{124}Sn, ^{128}Te, ^{130}Te, ^{136}Xe, ^{148}Nd, ^{150}Nd, ^{154}Sm, ^{160}Gd, and ^{198}Pt. Examples of DCE reactions relevant to $\beta^-\beta^-$ decay include (^{20}Ne,^{20}O) on ^{116}Cd, ^{130}Te, and ^{76}Ge targets, while those relevant to $\beta^+\beta^+$ decay include (^{18}O,^{18}Ne) on ^{48}Ti, ^{76}Se, and ^{116}Sn targets. The beams are delivered by the Superconducting Cyclotron (CS) at LNS-INFN, and the experiments utilize the large-acceptance MAGNEX magnetic spectrometer [10], equipped with Focal Plane Detectors (FPD) for ejectile detection and the newly developed G-NUMEN array positioned around the target for γ-ray detection.

A pioneering experimental campaign conducted by the NUMEN collaboration demonstrated that precise DCE cross sections can be measured even at very forward angles [11, 12]. The resulting high-resolution spectra in terms of energy, mass, and angular distribution allow for the clear identification of ground-state-to-ground-state transitions of interest [3, 4]. Beyond DCE reactions, simultaneous investigations of other reaction channels will facilitate the evaluation of competing multi-nucleon transfer processes. Moreover, this comprehensive approach will enable a deeper understanding of the reaction mechanisms and a constrained analysis of nuclear states relevant to both DCE reactions and $0\nu\beta\beta$ decay.

However, the DCE cross section represents only a small fraction of the total reaction cross section, making it particularly challenging to study isotopes with extremely low cross sections relevant to $0\nu\beta\beta$ decay. These cases require highly sensitive measurements to distinguish the DCE signal from background noise, which must be minimized. Additionally, cross section measurements must achieve the highest possible statistical significance.

To meet these stringent requirements, a major upgrade of the CS and its associated infrastructure has been undertaken, enabling the delivery of high-intensity beams with intensities two orders of magnitude higher than before. Moreover, an advanced experimental setup for high-sensitivity spectroscopy has been developed, capable of operating in a high-radiation environment. The details of this upgrade are outlined in the Technical Design Report of the NUMEN project and its subsequent updates [13–18].

2 Focal Plane Detectors

The FPD are employed to track the reaction ejectiles and to determine their atomic number (Z), isotopic species (A), and atomic charge state (q) [19]. The limitations of the previous FPD system [20, 21], which consisted of a low-pressure gas drift chamber with proportional wires for primary electron multiplication, followed by a stopping wall of silicon detectors, are mainly related to its maximum operational rate of a few kHz and the limited radiation hardness of the silicon detectors.

The new generation of FPDs is designed to withstand rates of up to 50 kHz/cm, with radiation hardness up to a fluence of 10^{11} ions/cm^2/year. Additionally, it must provide particle identification capabilities for ejectiles with $Z \sim 10$ and $A \sim 20$, a time resolution better than 2 ns, a double-hit event probability below 3% across the entire FPD, and the ability to operate in a low-pressure gas environment (tens of mbar).

To fulfill these demanding requirements, a gas tracker has been developed in which the traditional multiplication wires are replaced by Multi-Thick Gas Electron Multiplier foils, followed by a stopping telescope composed of large-area Silicon Carbide (SiC) detectors and Thallium-doped Cesium Iodide [CsI(Tl)] scintillators. Details on the mechanical design, dimensions, and materials of the new FPD system, which is currently under development, are provided in [14, 15, 17, 18]. Here, we summarize its main characteristics and present the results of selected test campaigns conducted to evaluate its performance.

2.1 Gas Tracker

The gas tracker is designed to provide the phase space parameters of the reaction ejectile at the focal plane: x_{foc}, y_{foc} with resolution below 0.6 mm, and θ_{foc}, ϕ_{foc} with resolution below 5 mrad. High-resolution measurements of these parameters are crucial for accurate particle trajectory reconstruction, allowing the determination of the momentum vector at the target position and, consequently, the particle scattering angle and energy.

The adopted solution consists of a proportional drift chamber with a sensitive volume of $1200\times116\times108$ mm^3, equipped with a thin Mylar entrance window and filled with isobutane gas. The electron multiplication stage is based on a three-layer Thick Gas Electron Multiplier (M-THGEM), where each layer consists of an insulating plane sandwiched between two conductive layers with a high density of microscopic holes, manufactured using a multilayer PCB technique (total thickness about 1 mm). Primary electrons, generated along the ion track, drift under a uniform electric field toward the M-THGEM, where they undergo avalanche multiplication due to the high field. The resulting avalanche electrons are collected on a segmented anode, while the bottom electrode of the M-THGEM collects the positive ions. The charge distribution on the segmented anode provides the horizontal position (x, z) and the track angle θ, while the vertical position y and the angle ϕ are derived from the electron drift time.

To optimize the gas tracker performance, different M-THGEM hole geometries, coupled with strip- and pad-segmented anodes, were tested on the first gas tracker prototype. The characterization was carried out by measuring the currents flowing through the electrodes, employing a picoammeter with 7 ADC channels (15 pA precision, 10 mV sensitivity), as a function of the applied voltages, gas pressure, and incident particle rate. These tests were done using α sources and ^{18}O ions at E = 270 MeV at the INFN-LNS Test Bench [22]. The results demonstrated that high gas gains of $\sim 10^5$ can be achieved even at low pressures ($\sim$ 10 mbar) with minimal ion backflow (IBF). The IBF, defined as the fraction of positive ions collected on the cathode relative to the total charge, is a critical parameter since excessive ion backflow can distort the electric field in the drift region, degrading position resolution and drift time measurements.

The best performance was obtained using a full-coverage THGEM foil configuration with holes arranged in a square pattern, coupled with an anode segmented into pads in the x-z plane, five rows along z, each with 60 pads in the x direction, as shown in **Figure 1A, B**. Based on these findings, a second prototype was developed featuring the above M-THGEM and anode design but scaled to one-fourth of the full tracker length along the dispersive x-direction (**Figure 1C**). The position resolution of the prototype was assessed using a collimated α source, yielding a preliminary FWHM resolution in the x-direction of less than 1 mm.

The prototype (**Figure 1C**) was recently tested at the Pelletron facility of the Institute of Physics, University of São Paulo, Brazil. This experiment aimed to simulate the NUMEN beam conditions in terms of energy loss within a low-pressure gas volume. The following beams were used: (i) ^{12}C at 45 MeV and 37.5 MeV and (ii) ^{7}Li at 32 MeV and 28 MeV. The

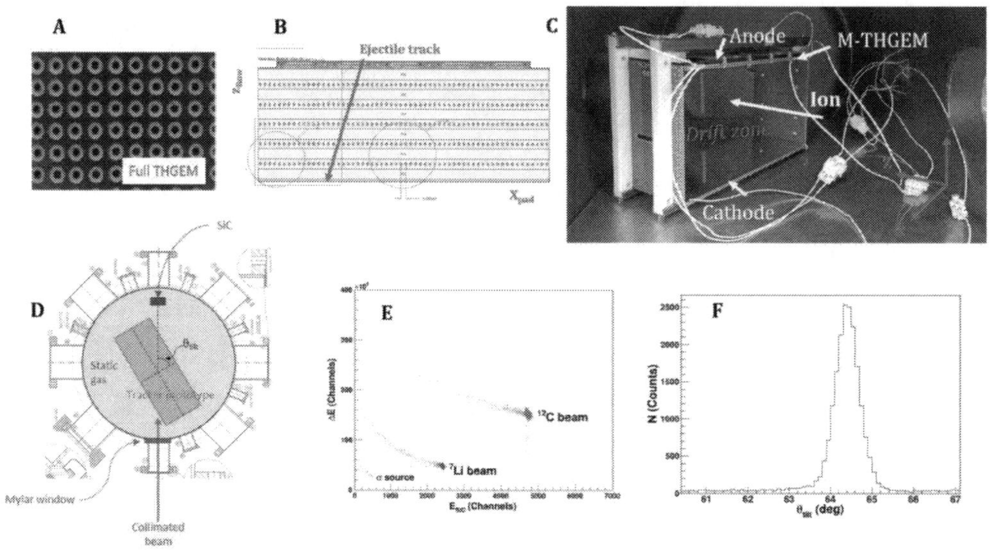

Figure 1. (A) Small area of the Full THGEM foil configuration. (B) Pad-segmented anode: 5 rows (z-direction), each with 60 pads in the dispersive x-direction. (C) Gas tracker prototype with the M-THGEM and anode shown in the same figure. (D) Schematic experimental setup. (E) Online bidimensional plot ΔE-E_{SiC} from the gas tracker prototype and the SiC detector. (F) Online reconstructed tilt angle θ_{tilt}.

experiment was carried out at the SAFIIRA facility, located at the 0° beamline, which is optimized for delivering high-quality beams into a multipurpose scattering chamber [23]. The beams were well-collimated (0.3 mm diameter, 0.2° FWHM divergence) and their intensity was carefully controlled (ranging from 10 Hz to 3 kHz). A schematic of the experimental setup is presented in **Figure 1D**. The accelerated beam passed through a thin Mylar window at the entrance of the 45 cm-diameter scattering chamber, filled with high-purity (99.95%) isobutane gas. The chamber pressure was varied across different runs. The tracker prototype was mounted inside the chamber at a tilt angle θ_{tilt}, which was varied from 40° to 60° in order to explore different ion trajectories. A SiC detector was placed downstream to stop the beam and provide a timing reference for measuring the electron drift time. During the experiment, the following aspects were investigated: (i) Tracker position and angular resolution; (ii) Dependence of performance on θ_{tilt} and (iii) Tracker response to beam rate variations. Data analysis is currently in progress, and results will be presented in a forthcoming publication. **Figure 1E** shows an online bidimensional spectrum of ΔE in the gas tracker *vs.* residual energy E_{SiC} in the SiC detector, while **Figure 1F** presents the online reconstruction of the tilt angle θ_{tilt}.

2.2 Particle Identification Wall

The particle identification (PID) is carried out by combining the measurement of the horizontal position x_{foc} at the focal plane, provided by the gas tracker, with the energy loss (ΔE) and residual energy (E_r) of the incident ions, detected by a dedicated array of 720 SiC-CsI(Tl) telescopes. These telescopes are arranged in 36 towers, each tilted by approximately 35° with respect to the vertical axis to partially compensate for the inclination of the focal plane relative to the optical axis of the MAGNEX dipole magnet.

Each SiC element has an active area of 1.5×1.5 cm^2, with a peripheral edge structure of 0.4 mm width, a thickness of 100 μm, and a 10 μm-thick dead layer on the back side. The

CsI(Tl) inorganic scintillator is 5 mm thick and is coupled to a Hamamatsu S3590 photodiode with an active area of 1×1 cm^2.

The first prototypes of such large-area SiC detectors were fabricated using epitaxial silicon carbide growth with two different doping concentrations, namely, 3×10^{13} atm/cm^3 and 9×10^{13} atm/cm^3, corresponding to different depletion voltages. These detectors were developed as part of the NUMEN R&D program to design a new PID Wall for the MAGNEX FPD. This marked the first time that SiC detectors with such a large area ($\sim$ 2.4 cm^2) and a thick epitaxial layer ($\sim$ 100 μm) were produced and characterized [24]. Doping concentration plays a critical role in the development of SiC detectors for NUMEN, as they must operate in a low-pressure environment. This requirement imposes a low depletion voltage to prevent discharge effects, necessitating a lower doping concentration than the conventional 10^{14} atm/cm^3.

To determine the full depletion voltage and doping profile of the SiC detectors, current-voltage and capacitance-voltage characteristics were investigated. Energy resolution and depletion depth were evaluated using radioactive α-sources. The tested detectors exhibited a good energy resolution ($\sim$ 0.5% FWHM), well within the NUMEN requirement of $\leq$ 2%. However, differences in depletion depth and doping profile were observed between detectors from the two wafers. The low-doping wafer exhibited a depletion depth lower than the nominal value and a poorer performance in terms of the doping profile.

Since a lower doping concentration is essential for NUMEN, new wafers were recently produced using a different doping process. The obtained doping profiles appear to meet the required uniformity standards, and further tests will be conducted once detector fabrication is completed.

Regarding radiation hardness, the PID Wall components have demonstrated excellent resistance to ionizing radiation. The SiC detectors have withstood fluences up to 10^{13} ions/cm^2 [14, 25, 26], while the CsI(Tl) scintillators have endured fluences exceeding 7.5×10^{11} ions/cm^2 [14], fully satisfying the NUMEN experiment requirements.

A test campaign was undertaken at the TANDEM+ALPI facility of INFN-LNL to evaluate the PID Wall's Z-identification capability, response to high rates and sensitivity to the tilt angle. A ^{16}O beam at 15 AMeV was used, with two PID Wall towers placed at forward angles inside the PISOLO scattering chamber. Preliminary results indicate: Z-resolution $\Delta Z/Z \sim 1/34$ and mass resolution $\Delta A/A \sim 1/47$, ensuring unambiguous particle identification; excellent time resolution in the order of hundreds of ps for the SiC detectors and double-hit probability below 2.5%, demonstrating the effectiveness of the chosen segmentation. Data analysis is ongoing, and the final results will be presented in a forthcoming publication.

Additional research on SiC sensors has recently been carried out at the Ruđer Bošković Institute, Zagreb. The focus of this study was to investigate the charge collection efficiency (CCE) across the entire active volume and near the detector edges, a crucial aspect for guaranteeing uniform response. This investigation was performed using the Ion Beam-Induced Charge technique, which employs focused MeV-range accelerated ions with a beam spot as small as $\sim 1\mu$m to probe charge transport within the detector. Preliminary results show excellent uniformity in the central region of the SiC sensors, where the CCE remains constant and a noticeable decrease in CCE near the sensor boundaries, where CCE tends to drop to zero. These findings are crucial for optimizing the detector design and ensuring a consistent response over the entire active area.

3 The G-NUMEN γ spectrometer array

DCE events can be detected with high selectivity using the MAGNEX spectrometer [27, 28], which provides an energy resolution of approximately 500 keV at low incident energy. This resolution is sufficient to distinguish the ground state from the first excited state for non-deformed nuclei in both projectile and target. However, for deformed target nuclei such as ^{110}Pd, ^{150}Nd, and ^{160}Gd, as well as for cases involving high incident energies, this resolution is inadequate. To address this limitation, the G-NUMEN γ-ray spectrometer was introduced around the target position. This system detects γ-rays in coincidence (or anti-coincidence) with projectile-like fragments in the FPD, allowing the discrimination of closely spaced energy states [13, 29]. G-NUMEN consists of 110 $LaBr_3$(Ce) scintillators coupled with Hamamatsu R6231 8-dynode-stage photomultiplier tubes (PMTs), with the entire detector system supplied by EPIC Crystal (EPIC from now on). Each crystal has a diameter of 38 mm and a length of 50 mm. Positioned 24 cm from the target, the detectors cover a solid angle of 20% and exhibit an expected total photopeak efficiency of 4% at 1.3 MeV.

The main requirements for the G-NUMEN detectors are [13]: (i) a time resolution of a few ns to differentiate between consecutive beam bunches and suppress random coincidence backgrounds; (ii) an energy resolution better than 10% at 200 keV; (iii) a low observational limit; (iv) stable PMT charge conversion gain (defined as the ratio of output pulse charge to incident photon energy) at count rates up to ~300 kHz; (v) a linear response of output pulse charge to incident photon energy and (vi) radiation hardness to fast neutrons up to 10^{11} neutrons/cm^2, the expected neutron fluence in a 1-month long experiment as estimated from FLUKA simulations [30].

The G-NUMEN detectors were tested under various count-rate conditions, replicating those expected in NUMEN experiments. The high count-rate tests involved the use of intense radioactive sources, a laser system delivering light to the PMT via an optical fiber, and a fusion-evaporation experiment conducted at the ALTO facility of the IJC Laboratory [31]. The results revealed that detector performance is influenced by count rate, high voltage (HV), and the electronic configuration of the PMT voltage divider.

Using the EPIC active voltage divider, the PMT charge conversion gain exhibited a dependence on the count rate, with the output signal degrading beyond a critical threshold that varied with the applied HV. This threshold defined the maximum sustainable count rate for the detector, which remained well within the operational range required for the NUMEN experiments. The percent variation of the PMT gain as a function of count rate at various HVs is presented in Figure 2A, where the limiting rate is identified as the point at which the gain undergoes a sudden increase. Furthermore, low-rate measurements showed that the charge of the output pulse deviated from linearity as the incident photon energy increased, with this effect becoming more pronounced at higher HV levels (Figure 2B).

To resolve these issues, a new active voltage divider was developed by the NUMEN collaboration, employing PMOS transistors to stabilize interstage voltages of the PMT dynodes. This configuration eliminates gain variation and count-rate limitations. Furthermore, optimizing the voltage gradient across dynodes (a "tapered" voltage divider) improves pulse linearity. With this new voltage divider a highly stable gain (within 0.5% across the full range), excellent energy resolution (2.8% FWHM at 662 keV, 6.4% at 122 keV), sub-ns timing resolution and an improved linearity with energy were achieved, fully meeting NUMEN requirements (figures 2D, E). A forthcoming paper will present a detailed analysis of these results.

A crucial experiment was performed at the ALTO facility (IJCLab, Orsay) to verify the capability of G-NUMEN (in coincidence with MAGNEX) to distinguish γ-rays from DCE events amidst intense backgrounds from projectile-target interactions. A ^{16}O beam at 83.5 MeV was directed onto a Ti target, followed by a Ta target. The experiment replicated the

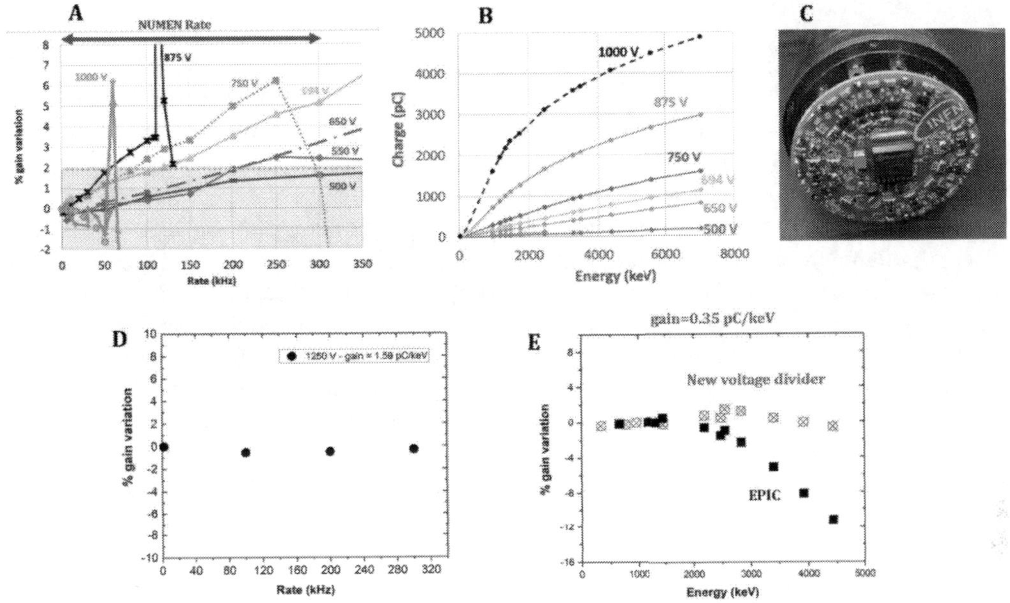

Figure 2. With the EPIC active voltage divider at various HVs (A) Percent variation of the charge conversion gain (normalized to the gain at low rate) *vs* count rate and (B) Charge of the output pulse *vs* incident photon energy (C) The newly developed active voltage divider (D) the same quantities shown in A for the new active voltage divider at HV=1250 V (E) percent variation of the charge conversion gain (normalized to the gain at low energy) *vs* incident photon energy with the two voltage dividers at the same gain, 0.35 pC/keV. The corresponding HVs are: 755 V for EPIC and 780 V for the new voltage divider.

signal-to-background ratio of a NUMEN experiment with a 10^{12} pps beam intensity. Preliminary results are consistent with previous GEANT simulations [13, 32], confirming that particle-γ coincidence allows for effective signal discrimination when high trigger selectivity is ensured, as in the case of MAGNEX for the NUMEN experiments. A complete analysis of the results will be presented in an upcoming publication.

While $LaBr_3$(Ce) scintillators are expected to be radiation-hard to fast neutrons, previous studies were limited to fluences of 10^8 neutrons/cm^2—three orders of magnitude below NUMEN requirements. To extend this study, the NUMEN collaboration carried out an experiment at the ALTO facility, exposing detectors to fluences up to 10^{10} neutrons/cm^2 using the LICORNE neutron source that provides fast neutrons (with energy up to 3 MeV) via the ^{7}Li(p,n)^{7}Be reaction at 16.7 MeV in inverse kinematics [33]. Measurements before, during, and after neutron exposure assessed changes in intrinsic background, time resolution, gain, and energy resolution. Preliminary results indicate an increased intrinsic background post-irradiation, with full recovery after one week. The light yield decreased by 4%, stabilizing at a 3% reduction after one week. The energy resolution degraded by 5% but fully recovered within 20 days, while the time resolution remained unaffected. These findings suggest that $LaBr_3$(Ce) meets NUMEN requirements up to 10^{10} neutrons/cm^2, though continuous energy calibration is needed to correct light yield variations. Future tests will explore neutron fluences up to 10^{11}-10^{12} neutrons/cm^2.

4 Target system

One of the most challenging aspects of the R&D activities within the NUMEN project concerns the production of targets composed of specific isotopes, which must be irradiated with intense ion beams while ensuring high energy resolution of the reaction products and allowing the collection of sufficient statistics for DCE reactions.

This can be achieved by depositing a few hundred nanometers of isotopic material onto high thermal conductivity backings characterized by low stopping power, high mechanical strength, high purity, and minimal thickness. Suitable substrates include graphite foils such as Highly Oriented Pyrolytic Graphite (HOPG) or alternative multi-layer graphene (MLG) candidates. These materials exhibit high in-plane thermal conductivity, enabling efficient heat dissipation to a cooled frame that acts as the mechanical support for the target. The developed cooling system consists of a target encased in a copper holder, mounted on top of a cryocooler capable of dissipating up to 20 W while maintaining a temperature of 40 K. Due to the high-radiation environment at the target position, an automated system was necessary to handle the replacement of irradiated targets and their storage, eliminating the need for human intervention. A detailed description of the entire target system and its automated handling is provided in [17, 18].

Initial target production attempts utilized HOPG substrates (supplied by Optigraph, Germany). These first depositions were performed using electron-beam evaporation of Ge, Te, Se, Mo, Sn, and Cd isotopes, varying the evaporation rate, modifying the thickness and temperature of the HOPG substrate, and, in some cases, employing a deposition buffer to enhance the diffusion of evaporated material onto the backing surface. To characterize the target and substrate properties, various analytical techniques were employed, including: (i) Transmission of α-particles from a radioactive source through the sample, followed by their detection using a 100 μm-thick silicon detector to assess thickness and non-uniformity (APT method); (ii) Scanning Electron Microscopy (SEM) to investigate surface morphology with nanometer-scale resolution in the planar axes and (iii) Rutherford Backscattering Spectroscopy (RBS) at INFN-LNL, using proton and α-particle beams to evaluate thickness and conduct elemental analysis of both target and substrate.

Target production proved to be highly complex due to the stringent requirements for low thickness and high uniformity, which are essential for preserving good energy resolution. For isotopes such as Ge and Te, the best results were obtained under standard evaporation conditions, i.e., without substrate heating, without a deposition buffer, and at a low evaporation rate (0.2–0.5 Å/s) [34]. However, successful Mo target depositions were achieved via electron-beam evaporation onto HOPG backings at a substrate temperature of 300°C, yielding excellent Mo adhesion [35]. For this isotope, substrate heating was found to be crucial in mitigating post-evaporation stresses and preventing delamination effects, which typically lead to significant non-uniformities on the target surface.

An important aspect of target R&D involves investigating radiation damage to the substrate, particularly the formation of defects such as atomic dislocations, vacancies, and broken chemical bonds. Understanding these degradation mechanisms helps to determine the replacement frequency of target-substrate systems during experiments, as crystalline defects impair the substrate's thermal conductivity. To evaluate radiation damage, HOPG foils were irradiated with monoenergetic 14 MeV neutrons from a Deuterium-Tritium neutron generator at the Instituto de Estudos Avançados (IEAv) in São José dos Campos, Brazil, to fluences of 1.7×10^{10} neutrons/cm^2. Structural analyses revealed a 50% reduction in thermal conductivity due to neutron-induced crystal defects. Morphological and crystallographic characterizations were performed using X-ray diffraction (XRD), micro-Raman spectroscopy, Atomic Force Microscopy, and SEM [36]. Simulations of 14 MeV neutron interactions with ^{12}C demonstrated that crystalline damage results from multiple physical mechanisms, including atomic displacement and energy transfer to both the crystalline lattice and ionization of HOPG. The dominant contribution to the damage stems from the interaction of α particles, produced in the ^{12}C(n,α)^{9}Be channel.

To evaluate the energy resolution (ΔE) of the first target prototypes, analytical calculations were carried out for ejectiles detected at 0° relative to the incident beam in the ^{130}Te(^{20}Ne,^{20}O)^{130}Xe DCE reaction at 15 AMeV. The analysis considered various contributing factors, including beam energy spread, the intrinsic resolution of the MAGNEX spectrometer, and target-substrate effects [34]. It was found that the non-uniformity of the HOPG substrate was the dominant contribution to ΔE. To ensure sufficient separation between the ground and first excited states of ^{130}Xe, the substrate's non-uniformity should be $\leq$10%. However, multiple HOPG samples exhibited higher non-uniformity, rendering them unsuitable as substrates for NUMEN targets.

Given the limitations of HOPG, current R&D efforts focus on exploring MLG foils as alternative target substrates. Various MLG samples, produced by different methods and sourced from multiple companies (Kaneka, ACF-metals, Nanotech, Micromatter), are being evaluated and characterized. Their thermal conductivity is estimated from the density, through XRD measurements, and literature-reported thermal diffusivity values. Thickness and non-uniformity are measured using APT, while radiation damage is investigated via Raman spectroscopy to examine defects in the graphene structure. Additionally, new synthesis techniques for MLG foils are being explored, including low-pressure filtration and pyrolysis of Kapton. The ultimate goal is to develop materials with tailored structural properties (e.g., high order, smooth surface, layered morphology, thermal stability) suitable for various applications, including NUMEN targets.

For NUMEN-relevant Xe and Kr targets, producing thin samples is particularly challenging due to their inability to form solid compounds. Initial implantation tests were conducted at IFUNAM (Mexico) using electromagnetically selected ^{127}I and ^{81}Br ions implanted into a 50 μm-thick graphite substrate at energies near 1 MeV (200–300 nm depth) [37]. These ions were chosen due to their similarity to noble gases. Post-implantation characterization was obtained via RBS using ^{1}H, ^{12}C, and ^{28}Si beams at various energies and scattering angles. The same methodology will be applied to produce noble gas-implanted Xe and Kr targets in future studies.

5 Conclusions

The NUMEN project pursues an ambitious program focused on the study of DCE reactions. Its primary goal is to precisely measure DCE cross sections at very forward angles, with the hope of gaining insights into the NMEs involved in $0\nu\beta\beta$ decay, which would be crucial for determining the effective neutrino mass should such a decay be observed.

Achieving an accurate measurement of DCE cross sections, which represent only a small fraction of the total reaction cross sections, has required an extensive upgrade of the CS at INFN-LNS and the MAGNEX spectrometer. This is essential for investigating all nuclear candidates for $0\nu\beta\beta$ decay. In particular, significant technological advancements have been necessary for the production of thin, highly uniform targets and substrates capable of withstanding intense ion beam irradiation. New substrate materials are currently being evaluated and characterized using various analytical techniques. Furthermore, considerable efforts have been dedicated to the development of advanced detectors and specialized electronics designed to operate in high-radiation environments.

The construction of the NUMEN experimental setup is at an advanced stage. Several campaigns aimed at characterizing the detectors and targets, as well as assessing their radiation hardness, have been successfully completed, while others are currently in progress. These efforts are crucial to preparing for the project's final phase—commissioning and the start of experimental runs.

References

[1] J. Engel and J. Menendez, Rep. Progr. Phys. **80** 046301 (2017)
[2] A. Belley, T. Miyagi, S. R. Stroberg, and J. D. Holt arXiv:2307.15156v1
[3] F. Cappuzzello et al., Eur. Phys. J. A **51** 145 (2015)
[4] F. Capuzzello et al., Eur. Phys. J. A **54** 72 (2018)
[5] F. Cappuzzello et al, Progr. in Part. and Nucl. Phys. **128** 103999 (2023)
[6] J.L. Ferreira et al., Phys. Rev. C **105** 014630 (2022)
[7] J. I. Bellone et al., Physics Letters B **807** 135528 (2020)
[8] H. Lenske, F. Cappuzzello, M. Cavallaro, M. Colonna, Prog. Part. and Nucl. Phys. **109** 103716 (2019)
[9] H. Lenske, J. Bellone, M. Colonna and D. Gambacurta, Universe **2024** 10, 202
[10] F. Cappuzzello, C. Agodi, D. Carbone and M. Cavallaro, Eur. Phys. J. A **52** 167 (2016)
[11] V. Soukeras et al., Results in Physics **28** 104691 (2021)
[12] C. Eke at al., Results in Physics **67** 108037 (2024)
[13] F. Cappuzzello et al., The NUMEN technical design report, Int. J. Modern Phy. A **Vol. 36** No.30 2130018 (2021)
[14] P. Finocchiaro et al., Universe **2020**, 6, 129
[15] F. Cappuzzello et al. Frontiers in Astronomy and Space Sciences, V.**8** 668587 (2021)
[16] C. Agodi et al., Universe **2021**, 7, 72
[17] D. Calvo et al., Journal of Physics: Conf. Ser. **2374** 012029 (2022) IOP Publishing
[18] D. Calvo et al., Nucl. Inst. Method in Phys. Res. A **1041** 167336 (2022)
[19] F. Cappuzzello et al., Nucl. Inst. Method in Phys. Res. A **621** 419 (2010)
[20] M. Cavallaro et al., Eur. Phys. J. A **48:59** (2012)
[21] D. Torresi et al., Nucl. Inst. Method in Phys. Res. A **989** 164918 (2021)
[22] I. Ciraldo et al., Nucl. Inst. Method in Phys. Res. A **1048** 167893 (2023)
[23] V.A.P. Aguiar et al., Rev. Sci. Instrum. **91** 053301 (2020)
[24] D. Carbone et al., Nucl. Inst. Method in Phys. Res. A **1069** 169960 (2024)
[25] S. Tudisco et al., Sensors **2018**, 18 2289
[26] C. Altana et al., Sensors **2023**, 23(14), 6522
[27] S. Calabrese et al., Nucl. Inst. Method in Phys. Res. A **980** 164500 (2020)
[28] M. Cavallaro et al., Nucl. Inst. Method in Phys. Res. B **463** 334 (2020)
[29] J.R.B. Oliveira et al., J. Phys. Conf. Ser. **1056**:012040 (2018)
[30] T. T. Böhlen, F. Cerutti, M. P. W. Chin et al., Nucl. Data Sheets **120** 211 (2014)
[31] E. M. Gandolfo et al., Instruments **2023**, 7, 28
[32] J.R.B. Oliveira et al., Eur. Phys. J. A **56** 153 (2020)
[33] E.M. Gandolfo, Ph. D. Thesis **2022**, Dip. Fisica "Ettore Pancini", Univ. di Napoli Federico II, Napoli, Italy
[34] M. Fisichella et al., Eur. Phys. J. Web of Conferences **285** 01001 (2023)
[35] M. Fisichella et al., to be published
[36] M.A. Guazzelli et al., Diamond & Related Materials **151** 111803 (2024)
[37] S. Sandoval-Hipólito et al., Nucl. Inst. Method in Phys. Res. B **555** 165485 (2024)

https://doi.org/10.1051/epjconf/202533303007

Reevaluation of Pu electrodeposits for measurement of environmental samples by Accelerator Mass Spectrometry

L.R. Romero[1], C. G. Méndez-García[2], P. Santa-Rita, E. Vázquez[3], and C. Solís[3].

[1]Facultad de Ciencias, Universidad Nacional Autónoma de México (UNAM), Mexico City, Mexico.

[2] Investigadora por México CONAHCyT - Instituto de Física, Universidad Nacional Autónoma de México (UNAM), Mexico City, Mexico.
[3]Instituto de Física, Universidad Nacional Autónoma de México (UNAM), Mexico City, Mexico.

Abstract. The different plutonium (Pu) isotopes have been present during the last 85 years due to nuclear events; given their low abundance and long half-life, the analysis of Pu is possible using the AMS technique. The present work analyzes the optimization of the Pu extraction protocol and the tuning of the system to open the line of study of this element in the LEMA by characterizing Pu alpha sources and their subsequent recovery for AMS measurement. The results show a chemical yield within a range of 38 -86% with the optimal extraction and electrodeposition conditions, while the recovery results indicate that it is possible to obtain up to 86.46% of the total electrodeposit for study by AMS.

1 Introduction

Plutonium (Pu) is a very important element in nuclear technology that poses significant challenges and opportunities for nuclear forensic science, a discipline that specializes in the identification of sources of radioactive emissions, applying advanced techniques to trace the origin of the material [1].

Among the plutonium radioisotopes ^{239}Pu, ^{240}Pu and ^{241}Pu are of particular interest due to their radiological relevance and low abundance, which makes their precise quantification through techniques such as Accelerator Mass Spectrometry (AMS) indispensable; however, an inherent challenge of the AMS technique is its inability to determine the efficiency of chemical processes directly during the extraction of these radioisotopes [2].

To overcome this obstacle and optimize accuracy in plutonium analysis, it is necessary to validate the efficiency of the extraction methodologies, this is carried out through the use of Reference Materials with Certified Values (RMC), allowing a quantitative evaluation of the protocol performance. Alpha Spectrometry (AS), in particular, presents itself as a valuable complementary technique, since it allows a direct correspondence between the count rate and the total activity of the sample. In addition, the electrodeposition technique minimizes analyte losses, thus facilitating reliable determination of chemical performance.

2 Methodology

The general protocol for the extraction of $^{239+240}Pu$ in the IAEA-410 and IAEA-300 certified reference materials is described in **Fig. 1**. The purpose of this protocol is to extract Pu from the samples for posterior electrodeposition and characterization by SEM-EDX. Once this process is completed, alpha spectrometric analysis is done to determine the best parameters for measurement by AMS.

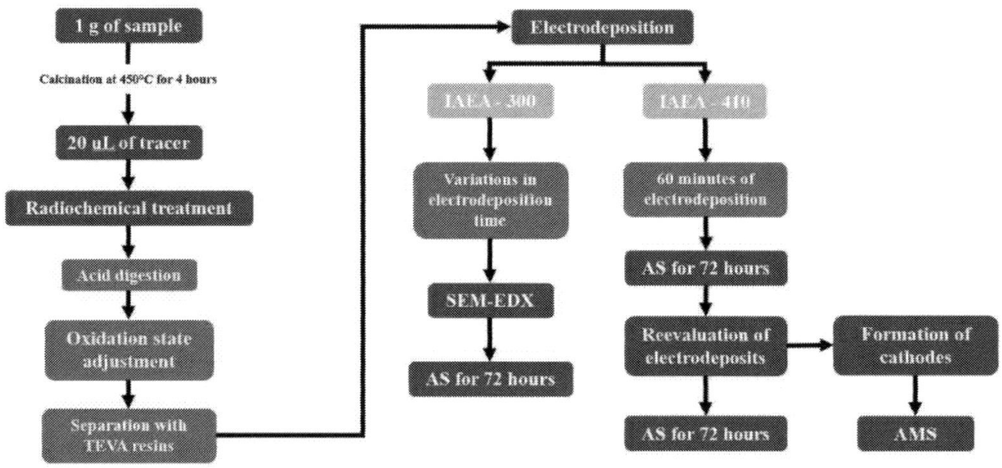

Fig. 1. Diagram of the methodology used for the study of $^{239+240}Pu$ by AMS.

The protocol used for the radiochemical treatment was based on the one provided by Elena Chamizo in her doctoral thesis *"Measurement of Plutonium isotopes, ^{239}Pu and ^{240}Pu, by mass spectrometry with low energy accelerators"* [3]. This starts with the calcination of 1 g samples at 450°C for 4 h, to later add 20 uL of ^{242}Pu tracer equivalent to (25.56 ± 2.52) mBq-mL^{-1}, with 5 mL of HCl 0.3 M.

In **acid digestion**, the resulting samples are evaporated at 80°C; once evaporated, 20 mL of HNO_3 was added and 3 mL of H_2O_2 was appended dropwise to avoid effervescence. This was repeated 4 times. Finally, 1 mL of 40 % HF was added, the result was centrifuged and the supernatant obtained was filtered. **The oxidation state adjustment** consists of three phases, *the reduction of the plutonium* by adding 1 mL of HNO_3 8 M together with 10 mL of H_2O; *the state adjustment* by dissolving 50 mg of $FeSO_4$ salt and waiting 15 minutes; finally, *the oxidation process* where 8 mL of HNO_3 8 M is added, then 100 mg of $NaNO_2$ salt and, in the same way, wait 15 minutes and add 3 g of $Al(NO_3)_3 \cdot 9H_2O$ salt, concluding with a dissolution at 0. 5 M.

For the **separation of Pu** a sequential extraction of actinides was performed using TEVA ion exchange resins (Eichrom Industries, Inc) which are conditioned with 10 mL of HNO_3 3M. The sequential extraction is given as follows: the sample obtained from the oxidation adjustment is passed through the resin, the U fraction is separated with 40 mL of HNO_3 2 M, the Th is eluted with 15 mL of HCl 6 M and finally, Pu fraction is obtained by adding 25 mL of a solution of HNO_3 0.02 M and HF 0.002 M, this extracts it and reduces its oxidation state to a trivalent state.

2.1 Method of electrodeposition

The method used for the electrodeposition was the one given by Lars Hallstadius in his article *"A method for the electrodeposition of actinides"* [4]. Once the Pu fraction is extracted, the dilution is evaporated to small volume, 1 mL of $NaSO_4$ 0.3 M is added and allowed to evaporate completely. The precipitate obtained from the evaporation is recovered with the addition of 300 µL of concentrated H_2SO_4 plus 4 mL of deionized H_2O, the dilution was transferred to the electrodeposition cells. The vial containing the dilution was washed again with 4 mL of deionized H_2O and, in the same way, it was transferred to the cell.

Subsequently, two drops of thymol blue indicator were added to the solution in the cell and the pH was adjusted to 2.5 using pure ammonia. During this adjustment, the indicator changed color from pink to yellow as the desired pH was reached. Electrodeposition was carried out for 60 min with a current of 1.2 A.

2.2 Reevaluation of the method for obtaining the electrodeposit

The electrodeposits were subjected to three different recovery methods, which are described below:

- **First method (A):** the planchettes were left in 20 mL of HNO_3 3 M with stirring overnight.
- **Second method (B):** The planchettes were immersed in 20 mL of HNO_3 (2.5 % v/v) and HCl (2.5 % v/v) with overnight agitation.
- **Third method (C):** The planchettes were left in 40 mL of HNO_3 (2.5 % v/v) and HCl (2.5 % v/v) with agitation for 1 h at 60°C.

After applying the methods, the planchette was washed with 10 mL of 5 % v/v HNO_3 and evaporated to dryness. Subsequently, the electrodeposition process was carried out for 60 min and the analysis by Alpha Spectrometry (AS) under the same conditions established in the initial study.

2.3 Formation of cathodes for AMS

Defined the method with the highest recovery, we proceed to make cathodes for posterior analysis by AMS. After the complete evaporation of the recovery dilution, the dried residue was recovered with 2 mL of HNO_3 1 M and transferred to a 15 mL centrifuge vial. The beaker containing the recovered sample was washed three times with 3 mL of deionized H_2O each time, reaching a total volume of 11 mL in the vial.

After obtaining the desired volume, 100 µL of Fe carrier was appended and the pH of the sample was adjusted to 9 to proceed with centrifugation. The resulting precipitate was lyophilized for 24 hours and later calcined at 800°C for 2 hours in porcelain crucibles. Finally, the residue obtained was mixed with 0.3 mg of powdered Al and pressed onto aluminum cathodes.

3 Results

The SEM images are shown in **Fig. 2**, where it is observed that, at shorter times, there is a low coverage, however, at 40 minutes an increase in coverage is observed, but with areas without deposits and, at longer times, the sample is in better conditions for its study, since it allows obtaining a homogeneous and well covered electrodeposit.

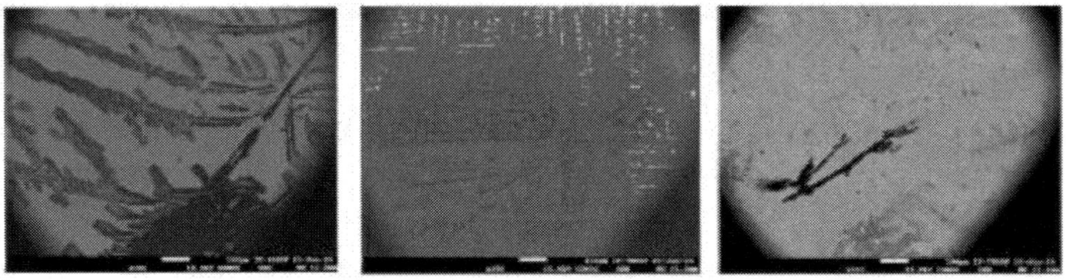

a) 20 minutes b) 40 minutes c) 60 minutes

Fig. 2. Images of electrodeposits by SEM-EDX.

The chemical yields obtained by alpha spectrometry show that group C has a higher average yield, reaching **86.5 %,** while those of group A achieved **37.7 %**. In contrast, group B samples had a considerably lower yield of only **2.6 %**. The original activity of the samples was calculated as a function of the chemical yield with respect to the reference value. The samples in the third group exhibit an activity of $^{239+240}Pu$ that closely approximates that reported in the RMC IAEA-410 certificate, which is **4.7±0.5 Bq-g-^{1}**. This result validates the efficacy of the methodology employed.

Of the three methods evaluated, the third method proved to be the most efficient, with an average recovery of **78.4 %**, followed by the first method, which obtained **60.7 %** and, finally, the second method with **8.8 %**, **Fig. 3** presents the above mentioned; however, in the last one, an increase in the activities after recovery was observed, probably due to variations in the electrodeposition process, which was carried out in alphabetically ordered series and could have included differences in factors such as time or current.

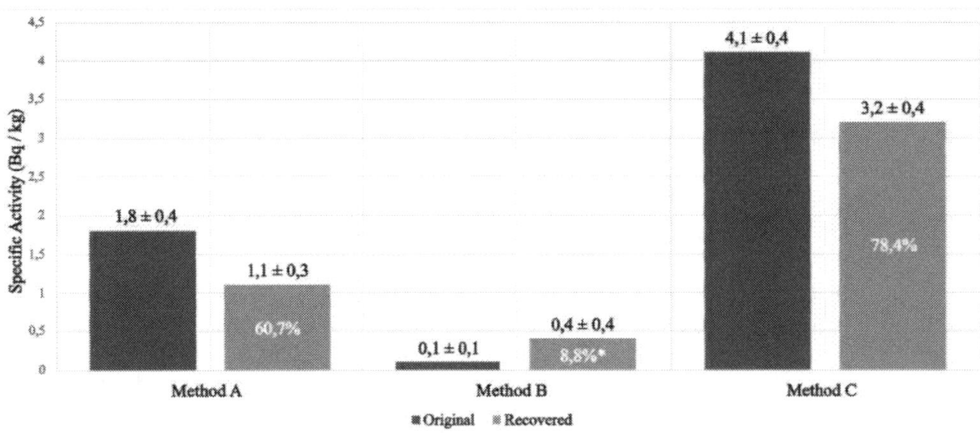

Fig. 3. Specific original and recovered activities with the three methods applied and their respective recovery percentages.

Before implementing the methodology described in this study to analyze $^{239 \text{ and } 240}Pu$ in environmental samples, a cross-validation between the AS and AMS techniques was carried out using IAEA-410 certified reference material. These show good agreement between the values obtained by AMS and those determined by AS as observed in **Fig. 4**, thus validating the reliability and accuracy of the proposed method.

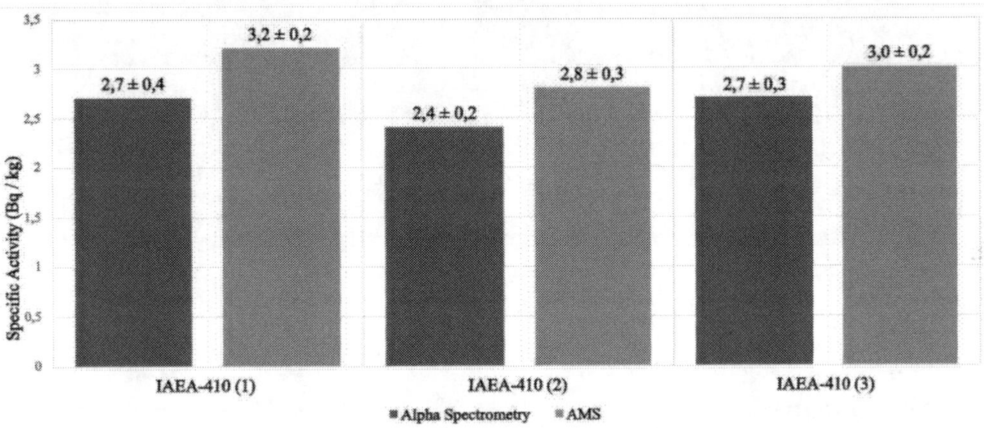

Fig. 4. Comparison between the specific activities of $^{239+240}Pu$ of RMC IAEA - 410 measured by AS and AMS.

Lake sediment samples at three different depths were processed using the same methodology applied to the RMCs. Results indicate variability in $^{239+240}Pu$ concentrations between samples, with values ranging from 2.2 mBq•g^{-1} in the Intermediate Segment to 3.4 mBq•g^{-1} in the Deep Segment, demonstrating differences in the accumulation or retention of these radioisotopes in the lake sediments analyzed.

The quantities recovered after treatment varied from 1.2 mBq•g^{-1} in the Deep Segment to 2.3 mBq•g^{-1} in the Intermediate Segment. The percent recovery (% REC) averages **64.7 %** with a standard deviation of 4.8 %. This indicates some variability in recovery efficiency between samples. The Intermediate Segment shows the highest recovery at **72.6 %,** while the Deep Segment has the lowest recovery **at 53.2 %.** The **Fig. 5** show the comparation between this last results.

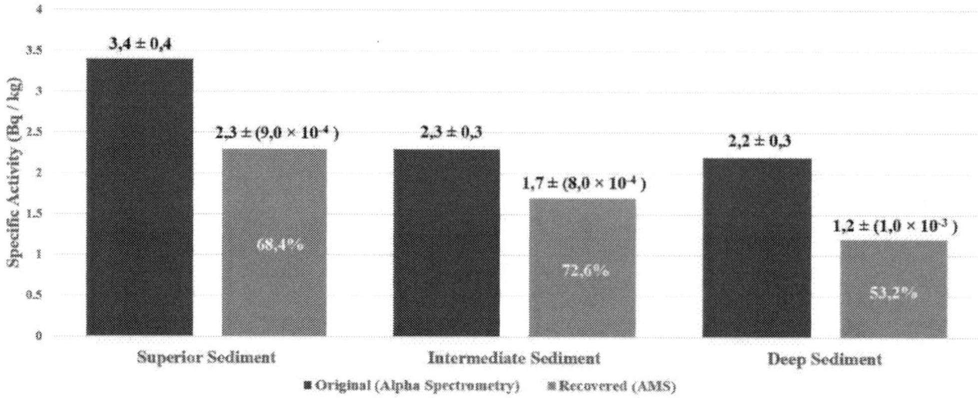

Fig. 5. Original (AS) and recovered (AMS) specific activities with the third method and their respective recovery percentages.

This suggests that the methodology applied is generally effective, but with variations in recovery efficiency between samples, which could be due to differences in treatment conditions or in the intrinsic characteristics of each sediment (matrix effect).

3 Conclusions

In this work, a protocol for the extraction and electrodeposition of Pu using IAEA-300 and IAEA-410 reference materials was validated. The optimal electrodeposition time to obtain a homogeneous and well coated electrodeposit was determined using SEM-EDX and EA. The optimized methodology allowed reevaluating Pu electrodeposits, implementing the third method for cathode formation for AMS. Demonstrating that this is suitable for the analysis of environmental Pu samples, both by AS and AMS.

This research can be considered as the first step for the opening of the Pu research line at LEMA, underlining the importance of optimizing the radiochemical process to obtain more accurate measurements in alpha spectrometry and AMS.

References

1. **Méndez-García C. Grisel, Romero-Guzmán E. Teresita, Hernández-Mendoza Héctor.** *La espectrometría de masas con aceleradores en la determinación de plutonio con aplicaciones en forense nuclear.* In: Espectrometría de Masas Con Aceleradores. Laboratorio Nacional. 1ra. Edición. Universidad Nacional Autónoma de México.; 2018:211-231.
2. **L.K. Fifield, R.G. Cresswell, M.L. di Tada, T.R. Ophel, J.P. Day, A.P. Clacher, S.J. King, and N.D. Priest.** *Accelerator mass spectrometry of plutonium isotopes.* Nuclear Instruments and Methods in Physics Research Section B: Beam Interactions with Materials and Atoms, 117(3):295–303, 1996
3. **Chamizo E.** *Medidas de Isótopos de Plutonio, 239Pu y 240Pu, Mediante Espectrometría de Masas Con Aceleradores de Baja Energía.* Universidad de Sevilla.; 2009.

4. **Fred Farris.** *Plutonium.* In Reference Module in Biomedical Sciences. Elsevier, 2023
5. **Lars Hallstadius.** *A method for the electrodeposition of actinides.* Nuclear Instruments and Methods in Physics Research, 223(2):266–267, 1984. 23
6. **Hernández, H.** *Desarrollo de nuevas técnicas para determinar actínidos en excretas por Espectrometría de Masas y Técnicas Radiométricas.* Universidad Complutense de Madrid.; 2011.

AUTHOR INDEX

EDP Sciences – Web of Conferences
17, Avenue du Hoggar
Parc d'Activité de Courtabouf
BP 112
F-91944 Les Ulis Cedex A
France

ISBN 979-8-3313-2496-4

CURRAN ASSOCIATES INC.
proceedings.com